Advance praise for
How to Be Animal

"With this book, Melanie Challenger fearlessly plunges into the biggest question of our time: How can we rediscover our animal selves, before it is too late? How can we discover our true place in the wider world we are destroying? Each of us has to answer this question for ourselves. This book is a guide for you on the journey."

—Paul Kingsnorth, author of *The Wake*

"Erudite, lyrical, delightfully troubling, and full of unexpected convergences. A wonderful exploration of the tensions that beset the human animal trying to find our way. I was entranced by this beautiful weave of history, biology, and philosophy."

—David George Haskell, author of *The Songs of Trees: Stories from Nature's Great Connectors*

"What an interesting book! The recognition that we are animals should come less as a slap in the face than as a welcome reminder of the great resources that can come from paying attention to the ways we and our various cousins handle our journeys on this difficult but beautiful planet."

—Bill McKibben, author of *Falter: Has the Human Game Begun to Play Itself Out?*

"Melanie Challenger's wonderful book teaches me this: our blazing continuity with the depth of time and the whole of life. It is a huge, complex, and triumphant thing: challenging, but also celebratory, courageous, mournful, and apprehensive. Her language is lovely: exact and lyrical and sparklingly full of suggestion and implication. It is a hymn to generosity. I know it will be something I will return to again and again."

—Adam Nicolson, author of *Sea Room: An Island Life in the Hebrides*

"Throughout our vexed shared history, animals have suffered from our insistence on comparing them to us, as if we were entirely separate organisms. Melanie Challenger's extraordinary, profound book turns the situation on its head. Perhaps only recently we have come to realize that this presumptive, if not arrogant, hierarchy does not exist. Only by seeing *ourselves* as animals are we going to survive. *How to Be Animal* is an utterly challenging, wholly essential book: it shows us how to be human."

—Philip Hoare, author of *The Whale: In Search of the Giants of the Sea*

"In *How to Be Animal*, Melanie Challenger offers a poetic and erudite meditation on the relationship of our species to the rest of the organic world, and especially to the species to which we are most closely related. Her compelling argument against human exceptionalism synthesizes the insights of scientists and philosophers from many times and places, and uses them, along with reflections on her own experiences, to analyze some of the most troublesome current political issues. She is particularly aware of the human desire for firm boundaries—between ourselves and other species, for example, or between our bodies and our minds—and she therefore stresses the elusiveness of such boundaries and the potentially devastating effects of our pursuit of them."

—Harriet Rivto, Arthur J. Conner Professor of History,
        Massachusetts Institute of Technology

"Melanie Challenger's erudite book confronts the refusal to embrace our animal nature and wrestles with the delusion and fear that underlie that refusal. In a time when so much of 'the rest' is collapsing, deconstructing the myth of human specialness, and the ways it is implicated in nonhuman suffering and destruction, is urgent. Challenger shows us that our moral awakening is not only about changing how we treat the Earth, but also about transforming how we see ourselves."

—Eileen Crist, author of *Abundant Earth: Toward an Ecological
        Civilization*

"Melanie Challenger's *How to Be Animal* will change the way you think. Every page has a mind-blowing moment or refreshing idea that deepened my understanding of the world and how we got to this moment in history. It's the most elegant and considered dismantling of the myth of human exceptionalism I've read. An illuminating, beautifully written, and unique philosophical inquiry by a wide-ranging and original thinker, and a powerful call for a new ethic for our relationship with the rest of the living world. I don't think there is a more important subject on earth today than getting to grips with how and why we do or don't value other beings, and Challenger excavates these questions with wisdom, compassion, rigor, and imagination. Quite simply, a rare and important marvel."

—Lucy Jones, author of *Losing Eden: Why Our Minds Need the Wild*

"A lyrical blend of timeless wisdom and truly new thinking. Seeing the world with animal eyes won't only change the way you think about animals—it will make you feel more human."

—J. B. MacKinnon, author of *The Once and Future World: Nature As It Was, As It Is, As It Could Be*

PENGUIN BOOKS

HOW TO BE ANIMAL

Melanie Challenger works as a researcher on the history of humanity and the natural world, and on environmental philosophy. She is the author of *On Extinction: How We Became Estranged from Nature*. Challenger received a Darwin Now Award for her research in Arctic Canada and the Arts Council International Fellowship with the British Antarctic Survey for her work on the history of whaling. She writes a blog called *Thinking with the Planet* and lives with her family in England.

How to Be Animal

A New History of
What It Means to Be Human

MELANIE CHALLENGER

PENGUIN BOOKS

PENGUIN BOOKS

An imprint of Penguin Random House LLC

penguinrandomhouse.com

First published in Great Britain by Canongate Books Ltd 2021
Published in Penguin Books 2021

LIBRARY OF CONGRESS CATALOGING-IN-PUBLICATION DATA
Names: Challenger, Melanie, author.
Title: How to be animal : a new history of what it means
to be human / Melanie Challenger.
Description: New York : Penguin Books, 2021. | Series: The indelible
stamp—The dream of greatness—The civil war of the mind—A stranger to
creation—The journey-work of the stars—Coda : on the loveliness of being
animal. | Includes bibliographical references and index.
Identifiers: LCCN 2020048166 (print) | LCCN 2020048167 (ebook) |
ISBN 9780143134350 (paperback) | ISBN 9780525506164 (ebook)
Subjects: LCSH: Human beings—Animal nature.
Classification: LCC GN280.7 .C43 2021 (print) | LCC GN280.7 (ebook) |
DDC 599.9—dc23
LC record available at https://lccn.loc.gov/2020048166
LC ebook record available at https://lccn.loc.gov/2020048167

Printed in the United States of America

Set in Garamond MT Std

But man can neither be understood nor saved alone.

Mary Midgley

CONTENTS

THE INDELIBLE STAMP

Man with all his noble qualities, with sympathy which feels for the most debased, with benevolence which extends not only to other men but to the humblest living creature, with his god-like intellect which has penetrated into the movements and constitution of the solar system – with all these exalted powers – Man still bears in his bodily frame the indelible stamp of his lowly origin.

Charles Darwin

The world is now dominated by an animal that doesn't think it's an animal. And the future is being imagined by an animal that doesn't want to be an animal. This matters. From the first flakes chipped from stone in the hands of walking apes at least several million years ago, history has arrived at a hairless primate with technologies that can alter the molecules of life.

These days, humans are agents of evolution with far greater powers than sexual selection or selective breeding. Thanks to breakthroughs in genomics and gene-editing technologies, the biology of animals, including humans, can be rewritten in various ways. We have created rodents with humanised livers or brains partly composed of human cells. We've made salmon that grow to our timetable. Scientists can sculpt DNA to drive lethal mutations throughout a whole population of wild animals.

Meanwhile, the rest of the living world is in crisis. In our oceans, our forests, our deserts and our plains, many other species are declining at unprecedented rates. In geological terms, we're an Ice Age, a huge metamorphic force. Our cities and industries have left their imprint in the soil, in the cells of deep-sea creatures, in the distant particles of the atmosphere. The trouble is we don't know the right way to behave towards life. This uncertainty exists in part because we can't decide how other life forms matter or even if they do.

All that humans have tended to agree on is that we are somehow exceptional. Humans have lived for centuries as if we're not animals. There's something extra about us that has unique value, whether it's rationality or consciousness. For religious societies, humans aren't animals but creatures with a soul. Supporters of secular creeds like humanism make much of their liberation from superstition. Yet the majority rely on species membership as if it is a magical boundary.

This move has always been beset by problems. But, as time has passed, it has become harder to justify. Most of us act according to intuitions or principles that human needs outrank those of any other living thing. But when we try to isolate something in the human animal and turn it into a person or a moral agent or a soul, we create difficulties for ourselves. We can end up with the mistaken belief that there is something non-biological about us that is ultimately good or important. And that has taken us to a point where some of us seek to live forever or enhance our minds or become machines.

None of this is to say that there aren't clear differences between us and everything else. Our conscious encounter with the world is a breathtaking fact of how life can evolve. We chat together about abstract concepts and chip images of ourselves out of rock. Like the beauty that a murmuration of starlings possesses, our experience seems to be more than the sum of our parts. From childhood onwards, we have a sense of identity, a kaleidoscope of memories. The sorts of skills and knowledge we bring into play in living and reproducing include the ability to fantasise and deceive, control certain urges and imagine the future. Through a blend of senses, emotions, hidden impulses and intimate narrative, we dream and we anticipate.

The human mind is an amazing natural phenomenon. Yet our kind of intelligence – having a subjective consciousness, among other things – does more than just enrich our experience of life. It provides far greater flexibility in our behaviour than might be possible without it, most especially with each other.

Little wonder then that we have spent much of history asserting that human experience has a meaning and value that is lacking in the rigid lives of other animals. Surely there is something about us that can't be reduced to simple animal stuff? Some might say that stripped of culture we become more

obviously akin to the other creatures on Earth, relying on wits
and body to get the energy to remain alive. Many works of art
have aimed to teach that lesson, needling the imagination with
the image of a human at the mercy of the forces of the natural
world. But even so, we recognise that this individual has a
potential for awareness that is unique in what we know – so
far – of life in the universe. Here we have it. The exhilarating
oddness of being something so obviously related to everything
around us, and yet so convincingly different.

We are the mythical being our ancestors once painted on
rock – a therianthrope, part animal, part god. There is the
animal body, the bit of us that bleeds and ages, and then there
is the exceptional bit that seems to come from our intelligence
and self-awareness, our spirit. As American political scientist
George Kateb has written, we are 'the only animal species that
is not only animal, the only species that is partly not natural'.
This idea can be found everywhere. We are animals as we
embrace and as our bloodied newborns slide from the bodies
of women but not when we make vows. We are animals as we
bite into the flesh of our meal but not in the workplace. We
are animals on the operating table but not when we speak of
justice. This split in the human condition, we are told, has not

only saved us from the meaningless lives of other creatures but forms the basis of the world we inhabit. It has raised us to the highest position in a hierarchy of life. It has left us with the impression that the human world is rich while the animal world is its pale shadow. And this has opened the way to a worldview in which our flourishing is the ultimate good.

It is, of course, perfectly possible to believe that humans are animals with no special origin or meaning, even peculiarly rapacious animals that the world would be better off without. But people rarely behave in accordance with this view – in other words, they usually continue to live as if the human world has meaning and rules of conduct that can be better or worse.

Perhaps it ought to end there. Yet we remain haunted. Many of our most common beliefs spring from an underlying refusal to accept that we are organic beings. Our kind of awareness has left us uncomfortable with the facts of an animal life. Animals suffer and die according to random events. Being a creature related to everything from an oak tree to a jellyfish brings with it threats like pathogens, injury, physical change and – for us – moral uncertainty. All that we love and value must be tugged out of an untamed landscape. This is both frightening and confusing. From this perspective, being animal is an embarrassment. Worse still, it is a danger.

Yet history has given us hope that we are different from the rest of the earthly rabble. What we truly are will save us from the fate of animals. Where other animals must suffer and perish, we have the gift of deliverance, whether into heaven, a glorious future or even merging with machines. We can be more than our animal bodies or our organic nature. What is important about us is somehow protected from the natural forces over which we fear we have no power. But this creates a strange amnesia. In convincing ourselves that there's a real and radical dividing line between us and all other organisms, we seem an impenetrable mystery.

Because of this, our relationship with being an animal is nothing less than bizarre. Most of us feel a frisson of anxiety that we live in a topsy-turvy world. Many of the things we most value – our relationships, the romantic sensations of attraction and love, pregnancy and childbirth, the pleasures of springtime, of eating a meal – are physical, largely unconscious and demonstrably animal. The things we most want to avoid – suffering, humiliation, loneliness, pain, disease, death – are born of animal instinct and the shared needs of an organism. Which is the truest part of the human experience, the animal, bodily feelings or the mental flickers of a wilful, storytelling intelligence? The trouble for us is that none of it quite makes sense. In our layered experience of the world, it's possible to believe we have left behind the blunt realities of being animal. Nothing could be further from the truth. Human life may be a blend of biology and dreams, but these dreams are still animal dreams. They are not separate to the bodies from which they arise. It's nonsense that our gifts have made us something that isn't animal.

So it is that we live behind a hidden membrane through which – at any moment – one of us may tumble to find ourselves on the other side. Opening our eyes, we face the truth of what we are, a thinking and feeling colony of energy and matter wrapped in precious flesh that prickles when it's cold or in love. We are a creature of organic substance and electricity that can be eaten, injured and dissipated back into the enigmatic physics of the universe. The truth is that being human is being animal. This is a difficult thing to admit if we are raised on a belief in our distinction.

What is different for our generation is that we now know

something that would have been blasphemous until very recently in human history. We know not only that the Earth is not the centre of the universe but that we are not the centre of life. Instead, we are an animal that finds itself aware of being an animal bound into the dark tissues of time and energy. The human species is an integrated part of the life on our planet, not an exceptional creation by itself.

If we had stayed in small bands on the African savannah, perhaps this knowledge would have had little consequence. As it is, there are now billions of us spread across all continents of the Earth. Nowadays, technological and industrial advances have distanced us from and, increasingly, medicalised our animal nature such that some of us treat our bodies as a malfunctioning part of us. The ongoing truth of our state can come as a shock. We are surprised by our frail flesh, the susceptibility of our bodies to desire as much as to disease. We spend millions to slow the ageing process, even more on the battle with ill health, and we are living through a determined project to remove reproduction from the messy chaos of our bedrooms and a mother's womb.

In our current industrial revolution, we have turned to the engineering of life in our pursuit of human wellbeing. It's hard to overstate the significance of this. Technologies that target our biology are constant reminders that we are animals. This is a problem for those who don't want to be one. A technological revolution that exploits the anatomy, physiology and behaviours of living organisms may be incompatible with human psychology. What we risk is a runaway process where our fear of being animal causes us to hammer out a more frightening world – not frightening in the sense that the world is nastier or more violent, but in a paradoxical reliance on technologies that aggravate the existential fears beneath us.

There's every reason to believe that when faced with a threatening reality, we will seek greater separation between us and the rest of nature. What form this separation might take is uncertain. One choice might be to do away with other animals or bring more of them out of a wild state and into submission. An easier course of action could be to put even greater accent on human exceptionalism, either by trying to make us superhuman or by shoring up our comforting beliefs. But yet another possibility is to do away with humans instead. It's easy to dismiss any one of those choices as overblown until we look around us. Each of them is being actively explored.

Of course, it's tempting to think that the uneasy relationship humans have with being animal is but an invention of modern civilisations or of a narrow band of philosophers. After all, the Lakota prayer *Mitákuye Oyás'iŋ*, most often translated as 'all are connected', is importantly different to that of the Roman Catholic catechism of the human person as *Imago Dei*. Some cultures amplify human distinction more than others. In light of this, some of the general statements in this book will refer more to the ideas of those civilisations. But the struggle with being animal isn't only a figment of culture. Our ideas are shaped around the anvil of human nature. These days, some say there's no such thing as human nature. But that is true only up to a point. Many things in our world work because we are – for want of a better description – a group of animals similar enough to be called a species. Our ideas about ourselves matter enormously, but they don't leave us untouched by common biological or psychological characteristics. The diverse ideologies of the world have attempted to solve some of the troubles that can come from being animals. But these are not just problems that evolutionary history creates for our redemption. We are a species that ponders its own condition. The underlying difficulties of being an animal remain, no matter what culture or era a person is born within. We all face real worries and dilemmas as a consequence of being alive among a multitude of lives.

This book is a defence of what it means to be an animal. It doesn't involve belittling us or losing sight of the obvious differences that mark us out. Nor does it result in a confused preference for what might be thought of as natural. Rather, this is an argument for a deeper understanding of how we think about life. Our animal origin is the story of our place in the world. It's the basis of how we give meaning to our existence. This is an impossible task without first accepting that humans are animals. This should be straightforward, yet it isn't. In truth, we live inside a paradox: it's blindingly obvious that we're animals and yet some part of us doesn't believe it. It's important to try to make some sense of this. And then, once we accept that we're animals, to think about what flows from that.

In a poem written in 1980, Galway Kinnell writes of how living things must contain within them a kind of self-love for their own unique biological form. In a way, this is the principle of survival. But he recognises that 'sometimes it is necessary/ to reteach a thing its loveliness'. What follows is an attempt to make sense of the kind of being that we are. Yet it's more than that. It's an invitation to refresh in our minds the loveliness of being animal.

THE DREAM OF GREATNESS

And yet is not mankind itself, pushing on its blind way, driven by a dream of its greatness and its power upon the dark paths of excessive cruelty and excessive devotion. And what is the pursuit of truth, after all?

Joseph Conrad

Falling upwards

Humans are part of a long process of emerging life, one that binds us to everything we see around us. 'From so simple a beginning,' declared Charles Darwin at the close of *On the Origin of Species*, 'endless forms most beautiful and most wonderful have been, and are being, evolved.' We don't yet know how the first living cells got going in the early stages of the Earth's history. Our world was then a rugged, mineral place, without hunger or judgement or any of the outrageous colours of a land with grasses and flowers. It is worthwhile to imagine standing in this smoking world, slammed by asteroids, and unthink the emergence of life. Somehow, in the heat of the deep-sea vents or in the shallow pools of that rough, smoking surface, primitive cells began to stir and gather through the peculiar business of energy conservation and flow.

'Life is, in effect, a side-reaction of an energy-harnessing reaction,' says biochemist Nick Lane. Or, as Austrian physicist Erwin Schrödinger put it in a series of public lectures he delivered in 1943 at the same time as the bloodiest battle in wartime history was coming to its close in Stalingrad, living matter seems to avoid 'the rapid decay into the inert state of "equilibrium"'. Whether or not we think of this kind of chemical event as rare or inevitable, we can identify it as one of the essential things that separates the living from the non-living.

Just as all life as we understand it persists by drawing on its environment – be it the barium-rich waters of a hydrothermal vent or the inside of an animal's cell – all known forms of life on Earth have the same rudimentary biochemistry. Life also shares heredity, which is to say that we can distinguish the glistening liveliness of a cresting wave from the organisms that might be carried in its waters because, while both require energy for their form, only life makes a child in likeness to its parent. Whether *E. coli* or an elephant, new life is generated from the divisions of a single cell. What is more, all living cells on our planet store the details of inheritance in deoxyribonucleic acid and involve certain chemical reactions accelerated by ribonucleic acid molecules.

More than three billion years ago these protocells likely became the first kind of bacterial life on Earth. Long before animal eyes could make sense of the landscape before them, the Earth's oceans were a vast bacterial kingdom. In time, evolution generated a fascinating change: pillars of rock colonised by cyanobacteria, tiny strands of bluish living entities that were doing something that would change the world – exploiting sunlight to stimulate their life cycle, producing oxygen in return. As these communities of bacteria grew, the cumulative effects of their presence made possible photosynthesising plants and the lungs of mammals like us, while limiting the opportunities for others like the beautiful *Spinoloricus cinziae*, an animal discovered only a few years ago in the Mediterranean Sea that is adapted to a life without any oxygen at all.

In 1967, American evolutionary biologist Lynn Margulis advanced the idea that animals and plants, as distinct from the first bacterial forms of life, owe much of their origins to an event called endosymbiosis. This is a process whereby one cell consumes another but without digesting it. There was considerable resistance to Margulis's theory, and it took corroboration through genetic studies more than a decade after she published her work before the theory became orthodoxy. The evidence for endosymbiosis is the presence, inside animal cells, of mitochondria and chloroplasts that divide independently and have their own DNA. Our mitochondria, absorbing nutrients and reservicing them as energy, are the ghosts of bacteria that our ancestors once ingested.

Imagine again standing in the same spot as in the smoking, lifeless Earth but that now it is the Cambrian era, around five hundred million years ago, and the animals have arrived. The seas contain creatures like the anomalocaridids, ornate, shrimp-like animals with two curled appendages for spooning other animals into their mouths. This is the moment most of the major animal phyla appear in the fossil record, followed by a massive phase of diversification. One theory for this explosion of life is that free oxygen was less limited. Newer research posits that there was a surge in calcium concentrations in the water. Still others hypothesise an arms race between predator and prey, and the evolution of eyesight. Nobody is entirely sure.

Yet the extraordinary diversity of life forms that we find in the Burgess Shale – the mineralised skeletons, the male and female anatomies of many of the species, the barbed or grasping shapes of the hunter and the hunted – show us how ensnared we are in a vast system of energetic interactions. The usual state of affairs is for living beings and the environments in which they survive to be subject to change and to death. Life forms may temporarily resist, but they can't do this forever.

NASA scientist Michael Russell, who has a beautiful copy of the *Great Wave off Kanagawa* on the wall of his office, once counselled me to remember that life is 'an entropy generator'. In other words, life decreases its internal entropy by using free energy from its surroundings and dissipating this as heat that, in turn, generates greater entropy in its environment. Entropy is a measure of how dispersed energy is among the particles in a system. For anyone who is not literate in physics, this can mean very little. Paul Simon is instructive in this situation. As he puts it in a song from 1972, 'Everything put together/sooner or later falls apart . . .' Think of a gin and tonic. The frozen water of the ice cubes has a lower entropy than the spiritous sea that surrounds it. The atoms in the gin are free to slosh about and take on the shape of whatever glass they're poured into, but the atoms in the ice are less randomly arranged and only lose their shape when energy in the form of heat loosens them up.

The ordered appearance of a life form can be understood as a temporary state of low entropy created by finding and using energy. In physics, eating other animals makes perfect sense once enough of them are swimming around. In a fun paper written by biologist Alexander Schreiber in an effort to counter the misconceptions of anti-evolutionists, he summarises the exchanges of energy and waste among animals and the environment: 'all organisms maintain their low entropy status by "eating" free energy and "pooping" entropy'. While energy is needed for the regulatory processes of our bodies, disorder must be exported in the wastes that we separate and expel. Perhaps even consciousness, Russell speculates, may be a means of 'using up excess energy'.

Meanwhile, the physicist Jeremy England has argued recently that reproduction in organisms is 'a great way of dissipating [energy]'. He has devised the theory that the second law of thermodynamics, or the law of increasing entropy, might cause matter to organise in lifelike ways. Should it turn out to be true, it will expose an underlying, shared process between life and non-life, a curious commonality between snow leopards and snowflakes. 'It is very tempting to speculate about what phenomena in nature we can now fit under this big tent of dissipation-driven adaptive organization.'

Life on our planet can be arranged, more or less, into autotrophs and heterotrophs, organisms that exploit energy from the sun or chemical reactions, and organisms that take energy from those who've already captured it. What is unusual about our species is that we've been able to use more and more energy without having to evolve into a different species. We've achieved this through a combination of social learning, complex culture, and technologies. We don't have to speciate to gain the claws of an allosaurus; we can share information to design a warhead or a power station. In other words, we change our tools rather than our bodies. Fire and spears did the trick for hundreds of thousands of years, until we devised the domestication of our food sources. The next big shift came in the mechanisation of processes that gave us the Industrial Revolution. This enabled us to draw ancient deposits of organic energy out of the Earth and burn them.

It took most of human history to reach a population of one billion. Immediately after the onset of the first Industrial Revolution, our global population grew by more than 50 per cent. Agricultural production doubled in the hundred years leading up to 1920. After 1920, it doubled closer to every decade. By the latter half of the twentieth century, our population began growing by around a billion every ten or fifteen years. The major limits on the growth of groups of living beings are usually food availability, competition, predators and pathogens. But our rising populations led to a scientific renaissance. Over the past hundred years or so we've uncovered some incredible methods for extending our lifespans and limiting the dangers of things like disease. Historians have nicknamed this era the Great Acceleration. It has given us everything from antibiotics to gene editing.

But as our populations and needs increase, so do our effects on crucial aspects of the Earth's systems. Much of this is common knowledge. Today the world is loosely divided into pessimists, who see an inevitable collapse on the horizon; optimists, who believe things will stabilise and that we will use reason to reshape a more sustainable world; and futurists, who don't like the look of either scenario, and seek investment for our escape. So it is that some of us are clanging the warning bells, some are designing clean energy, and the rest are trying to move to Mars. These are our times.

One thing we can be certain of is that our planet will one day be lifeless again. If we dream with the Earth for a moment, we can see that mass extinctions of life forms have happened occasionally, mostly due to comets and asteroids or Ice Ages caused by axial tilt, with carnivalesque consequences for the next generation of creatures. But it will not always be so. Hundreds of millions of years into the future, the higher quantities of solar radiation will have precipitated genetic and structural changes that will restrict photosynthesis. From a wild medley of plants like the ancient *Nothia aphylla*, stamped into the Rhynie chert in northern Scotland, there will be a shrinking stock of plant life and, eventually, none. The death of plants will ensure the end of much other life. We tend to be reckless or forgetful of plants, but they ground the potential for multicellular animals like us. Without them, we're lost.

Spin out to a few billion years from now, and the dynamic spiralling, convecting currents in the Earth's core, roiling molten forms of the elements that gave us the Iron Age and a coin imprinted with the words *In God We Trust*, will stop driving the magnetic field. With this will come the loss of those protective mechanisms we take for granted, along with the withering effects of solar wind. Our atmosphere of sunny days and breezy autumn mornings will be destroyed. Our oceans will evaporate. Before this perhaps there might have been some hardy survivors like *Deinococcus radiodurans*, nicknamed 'Conan the Bacterium', capable of enduring the most aggressive of conditions. But over greater timescales the surface of the Earth will melt, and all life will be at an end.

Still, to this day, none of us truly knows why we are here or what lies at the heart of the cosmos. One thing we do know is that living things and the forces that help bring them into existence can be nasty, even catastrophic. In one sense, biology seems to come out of violence, whether that is about energy or extreme heat. The solar system and the Earth's environments commit horrors on animals, even as they allow for new diversity and change. The village-sized comet that created the Chicxulub crater and temporarily reduced biodiversity on Earth also made way for new creatures, including us. From one point of view the asteroid is an evildoer, but from ours it is a godsend.

And yet we live according to rules about what is good and right. We feel strongly that our love and experience have value. Few of us seriously wish to challenge such a view. Yet it remains difficult to ground the significance of who we are and what we do when there doesn't seem to be anything straightforwardly good about the world from which we're born. When we look out there at the trees twisting in the wind, their leaves freckled with fungi, a bird nearby smashing the shell of a snail to get at the soft creature inside, we can struggle to find reasons for the meaning we give to life. When we try to find some kind of fundamental goodness, we come up against viruses, bacteria, pests and predation that challenge this at a profound level, even though they can lead to outcomes that, from our perspective, are positive. How much are we a part of all this?

The fact is that the moods and sensations that bring such intensity to our experience come from the raw materials of the world before us. We find the gentle dance of the trees beautiful and the meat of the bird might be delicious to us. The provenance of all we enjoy is a sequence of events and processes that have no obvious concern for what we now think of as the rightness or wrongness of something. Whether we like it or not, what we think is important in life seems to originate in a world that lacks any obvious moral sense.

Last summer I visited a dig in the Utah desert, where dinosaur bones were being worked out of the riverbeds along which the animals they belonged to once lumbered, searching for food and opportunity. There is an important mental shift that happens when dinosaurs are no longer curated objects in museums. One femur the palaeontologist was slowly scraping free stood taller than my child. But it wasn't only the legs that were massive. What was also laid bare was the enormous size of the teeth and the claws, the practical viciousness of them. This, I thought to myself, is what the need for energy can give you.

Shortly before we left, the young man showing us around pointed to some darker lines in the rocks across the way. 'We're not sure yet,' he said, 'but they could be evidence of early mammals.' My husband and I squinted at the cindery rocks – the burrows, perhaps, of the small, night-dwelling insectivores that might have become the class of warm-blooded, milk-bearing animals to which we belong.

Underlying arrangements of energy can push matters temporarily in the direction of expert killers like *T. rex* or favour superorganisms like ants. Either way, there are forces at work that don't suggest an onward movement or the inevitability of kindness. There's at least as much suffering as pleasure and as much distress as tenderness in the evolution of life. In the wild, where brains need to be a touch larger than among domesticated creatures, the dance of hunter and prey is a ceaseless fount of innovation. The ratcheting of intelligence or behaviour in the wild is often down to death-dealing of one kind or another. It is a stark truth that the majority of the newborn of other animals will not survive their first year of life. Evolution is sensible enough to spare the mother salmon any lasting attachment to the hundreds of eggs she may lay in each clutch, only 2 per cent of which will reach adulthood. It is better not to know what comes of the remaining 490 or so out of around five hundred that will end up in the stomachs of other creatures, even of other salmon, and on the blinis of humans at cocktail parties.

Yet the suffering and death that predators bring can be supportive of the overall ecosystem, with its abundance and richness of species. Predation exists at all levels of a natural system so enormous and vigorous that it's almost impossible to single out one animal or predict its effects. Predation can alter everything from disease dynamics in populations of animals to the sequestration of carbon. The classic example of the general effects of predators was the reintroduction of wolves to Yellowstone Park in 1995. The presence of the wolves set rolling a sequence of changes that are still making themselves known today. The last wolves of the area were killed by trappers in the 1930s. In their absence, numbers of elk increased. Once the wolves were back, the elk started moving around more, willow stands grew back that had been intensely grazed, restoring a food source for beavers, who spread and altered the stream dynamics, inadvertently providing habitats for fish and songbirds. And wolf kills also began to supply extra food for a host of scavengers, from ravens to grizzly bears. From the measure of biodiversity, the wolves are necessary and good. But that doesn't make predation a pleasant event in the lives of other animals. The research of Liana Zanette and Michael Clinchy has demonstrated that exposure to predators activates the fear circuitry in the bodies of those being hunted in ways still measurable weeks later. Stress of this kind must be paid for in the overall wellbeing of the animal.

We are born of this logic along with every other living thing on Earth. In most of our fundamental activities, we behave little different from other animals. We frequently kill and eat other animals for energy. We expel the waste products of our body in faeces and urine. We smell out chemicals exuded from each other's flesh and act in response. We communicate, seek mates and raise our offspring. One day something ends the vital processes of our bodies and they decay to almost nothing in a microbial tide. In this we join the rest of the animals. We are of the same wellspring. Our bodies and experiences are part of the precarious establishment of life that has given us frightening mechanisms and tactics as often as beneficent ones. And, taken as a whole, we too have brought at least as much pain and suffering to the Earth's disco of life as any other predator. And we too must feel pain and suffer.

Something that people often misunderstand is that evolution doesn't involve a direction. Think of the multicellular forms of plants, fungi and animals. They have evolved on a number of separate occasions, but there have also been reversions back to unicellularity, especially across fungi. One theory for how we have multicellular organisms is that different species of single-cell organisms cooperated. This is an appealing idea. But we can also consider the possibility that the division of a unicellular organism encountered some kind of failure. The cells that didn't separate later developed into individual tissues. Then there is the uncertain role of viruses. Now the bane of our lives and the source of illness and death, there's trace evidence that the genes of some viruses were involved in the ability of different cells to grow into the separate organs and tissues of an animal's body. The point here is that life is neither straightforwardly good nor progressive.

This might be unbearable if it weren't for the possibility that humans are special. Something important saves us from the threatening parts of earthly life. We're told this comes from the heart of human nature, some essential part of us. We can't see it and we can't measure it, but it marks us out as the most important life form on Earth. For those with deep belief in a creator, we have a soul that is unique to the human body. For secular humanist thinkers, we have soul-like mental powers that are unique to the human brain. Each is a reason for why humans aren't truly animals. At least, not in any crucial way.

These in turn are only echoes of much older disagreements about living things. Think of those philosophers and writers loosely grouped together as mechanists, who believed that the soul was in charge of the rational activity of the mind and vaguely attached to an otherwise machine-like body. They were at odds with animists, who saw all living matter as having *anima* – a soul – and the vitalists, who saw living matter as infused with a special essence that permitted movement and change.

For those who believe we were created by supernatural force, we are told 'the true nature of Man is that he is not an animal, but a human being made in the image and likeness of God', as written by the International Committee on Human Dignity, a more recent Catholic organisation aiming to reassert traditional scriptural views. In the humanist thought that emerged in Renaissance Europe, we're told that when we become a person, we are no longer simply a biological organism. We are elevated 'above all the other beings on earth', as Immanuel Kant put it. His is only one of any number of variations on a similar theme. In the words of American philosopher Eric T. Olson, a notable proponent of 'animalism', a branch of philosophy that argues the opposite, 'these philosophers are saying that each of us is a non-animal that relates in some intimate way to an animal'.

This idea – or something like it – has been a source of

profound solace for countless human lives. Whether it's the soul or some other property, these are ways to take humans out of a bewilderingly anarchic nature. They are methods to save humans from the difficulties that nature's amorality presents to us. The idea may take on different hues in different times and places. But there's always something transcendent about humans that rescues us. In this way, the major theories about the significance of human lives have the distinct whiff of psychological necessity rather than rational clarity.

Those reacting against the individualism inherent in Western democracies have tended to over-romanticise the realities of human and animal relationships in other cultures. It's certainly true that humans are far from identical in their views. Indeed, some animistic traditions extend value and personhood into plants, as well as other animals. Today, many of these kinds of smaller societies face pressures and persecution. The complexities of their perceptions and practices are often obscured. On the other hand, although many of the larger non-Western traditions see both humans and animals as inheritors of a spiritual realm, even in a belief system like Buddhism, whose history has involved plenty of internecine war and animal consumption, rebirth into another animal isn't a cause for celebration. So, too, the image of the vegetarian followers of Hinduism is far from the reality of people's lives. In India, where 80 per cent of the population identify as Hindu, only around 20 per cent have a vegetarian diet. Nor are humans and other animals alike in their spiritual worth. In the Taittiriya Upanishad the god Shiva makes clear that humans are unique in their ability to act on knowledge.

In any case, for our increasingly connected populations and economies, some ideas about human life have become almost universal. Today, human dignity both as something we possess and as an exclusive set of guidelines for how we should behave has spread around the world. In discussing the landmark South African court case against capital punishment, *State v. Makwanyane*, judge Kate O'Regan noted that 'recognizing a right to dignity is an acknowledgement of the intrinsic worth of human beings'. The concept we now use was written into the psyches and legal instruments of European nations while families were still repatriating those who died in the Second World War. In the 1949 German constitution, it is stated that 'the dignity of man shall be inviolable'.

Yet a little background is edifying. In German, the word for 'dignity' is *Würde*, closer in English to 'worth'. Dignity itself comes from an ancient Roman concept for the influence, mastery and character that males – not females – gathered throughout their lifetime. Just as the word 'value' in English is a somewhat uncomfortable blend of reputation and moral integrity, concepts of worth and dignity have always been associated with status. The Old French word *value* as a social principle was loaned from the eighteenth-century aesthetic and monetary assessments of paintings and enters modern parlance around the time millions of Europe's young men were being blown to pieces in the clays of France and Belgium.

The modern concept of 'worth' as special or intrinsic value was once attached to the Old English notion of 'manworth', which ranked humans according to the price one might give to the lord if one of his men was killed. First used in the laws of Hloðhaere and Eadric, this blood money was worked out not only in terms of compensation to be paid for the loss of various parts of the body, the legs or hands and so forth, but also the social rank of an individual. A nobleman like a thane might result in 1,200 shillings in compensation. A ceorl, a free individual who wasn't considered nobility – which gave us the term 'churlish' – was worth a mere 200. There was little or no value to the life of a servant. When it comes to what matters, societies have always struggled to unpick what we prize from power, wealth and status.

But the real difficulty comes from globalising an idea that fails to make adequate sense of what we mean by goodness in the first place. How do we decide that we are a good life form? At present, humans have defiled their home so thoroughly that the World Health Organization attributes seven million premature deaths to air pollution each year. That's more than ten times the number of deaths caused by anopheles mosquitoes. Our destruction of habitats has been so extensive that a review published in 2020 revealed the loss of two-thirds of the planet's vertebrates, based on long-term trends in population numbers. We can quibble over the figures, but the fact is our destructiveness obscures what is good about us. It begins to look as though dignity or whatever notion of transcendence we tell ourselves we possess is good only at our convenience. Not only that, but it's only good occasionally, and often only for some of us. Given this background, how do we decide that anything we do is much different from a shark's ampullae of Lorenzini, the sense organs that allow it to feel the electrical field of its prey? Unique, maybe. But good?

Love like an orangutan

Of course, it's a cheering thought that only human tendencies have full moral status, but it's difficult to square with the fact that we're animals. For centuries now, secular and scientific thought has sought to out-climb the mire of an amoral Earth. Cognisant of the painful or distressing aspects of how living

things overcome the problems of survival, thinkers have tried
to turn our morality into an abstract. We are told we can't use
nature or natural traits to determine what is good. In philosophy,
this has become known as the 'naturalistic fallacy'. In essence,
the fallacy claims to show why it's hard, if not impossible, to
divine what we ought to do or value from how something is
in nature. This has given us our modern belief that moral ideas
are human achievements that exist beyond our biology. This
seems like common sense.

But the idea that we can approach the moral experience of
humans as an abstract that has no relationship to the fact that
we're animals suffers from severe problems in its internal logic.
The modern desire for a break, an unbreachable gap between
us and the rest of our planet's life, is ultimately a desire for a
psychological and moral boundary that can both satisfy us and
make sense of the world we want.

The fact is that much of what we value is bound up with
being animal. Think of the bond between parents and children.
It's true that we love our children because of who they are,
their unique identities, and not only because of the interplay
of hormones in our bodies. Love isn't just some chemical ruse.
For humans, it is natural for our love to involve history and
insight. But imagine what it would mean to take the love of a
child and turn it over to a machine, a bunch of algorithmic
content, the kind of assembly of images that our smartphones
come up with to summarise a year of our lives, rather than the
animal compulsions, reasons, motivations and feelings that get
love kick-started. The fact that we give each other love and
support is a condition not of our rationalising but of our
compulsions as animals.

Other animals may experience only the sensations of what
we call love, but how far can we stretch the argument that they

are without moral weight? Orangutans are remarkable for having a life history not dissimilar to our own, at least in terms of mothers and children. Their pregnancies last a little under nine months. Infants remain with their mother, clinging to her body, feeding on her milk, for the first few years. They stay close, learning from her, for upwards of ten years, and make visits to her even after they are independent. Some claim we mustn't call these behaviours love. Perhaps that is fair. But it defies belief that these instincts aren't accompanied by feelings. The question is: do feelings matter? Do orangutan females nurture their children at a cost to themselves for so many years because they love them? Or because their biology creates necessity? What we can say is that their biology has given them reasons to behave in this way. It is how their children will survive and thrive.

Some argue that the parent–child bond is actually a roadblock to moral behaviour, pushing us to favouritism, even tribalism. But, for humans, tending to our young is not only about their physical and mental wellbeing, it is also constrained by the energetic costs to the caregiver. It is a role still largely undertaken by women, and an activity of such profound value to society that it remains staggering that not only do we not pay our mothers but often penalise them for taking time off. Even the focused, personalised element to child-rearing has its own logic. It not only benefits the child's developing sense of self but it is also determined by the time and energy love requires, particularly of women. At least, that's how it is for mammals like us.

The ultimate question about what is right is often then not whether it's natural but whether it's beneficial. The complication for us is in making sense of what that means. When we speak of benefits and goods, we often include more complicated assumptions about the relative value of the recipients of these goods. People are usually certain of their point of view, but solid facts are harder to come by. Either way, an absolute break with us and other animals is difficult to justify. The claim is easy enough to accept when it comes to a stinging nettle or a fungus, but it feels more problematic when thinking of other animals. Once you arrive at a being like an elephant – large-brained, aware, sensitive to its offspring, very different in outward form to us but quite similar in many other ways – on what grounds should we be convinced of an absolute border? And yet most of us are convinced.

The problem for us is that it isn't true. Whether there's a creator or not isn't the issue. Evolution by common descent doesn't prove or disprove the presence of God. And nor is there anything wrong in recognising that our unique biology is important to us. Humans needn't be superior or supernatural to have unique needs. But none of this cancels out the feelings and needs of other animals. For those of us who wish to live without magic or myth, who pride themselves on the truths of science, our relationship to the rest of life on our planet is a problem we have not yet made sense of. The minds of other animals, despite clear evidence that we are surrounded by a multitude of alternative psychologies, continue to be put to one side. Those who wish to believe in the moral progress of human civilisation must recognise that our relationship to other animals still largely lingers in the cold.

Some argue that we have no duties to other life forms because they can't reciprocate our acts of compassion. Certainly, it's harder to care a great deal without a return. But this is an excuse, not a reason. The only way to avoid the spread of moral compassion into the lives of other animals is to come up with a radical lie. We convince ourselves that our beliefs have little to do with the rest of the living world. After all, animals don't have minds or intentions like ours. Yet relying on this old idea saddles us with a moral system that is inconsistent to a fault.

And the inconsistency is beginning to show. The difficulty lies in the fundamental refusal to accept that we, too, are animals and that being an animal matters to us. This presses on our generation in new and powerful ways. In seeking meaning, we've told ourselves that we are the only animal with true value. Yet our consciousness, an unacceptably animal affair, tolerates the hypocrisy only by denying we are animals. The tension is unresolved in our times. A belief that has buoyed civilisation for millennia has reached the end of its usefulness.

A millennium of soul-searching

There's no doubt that humans are highly unusual. Our lives are unmistakably different from those of any other species. Flicking through the internet and watching films of people bounding about on the surface of the Moon or debating global security in the general assembly of the United Nations, this much is obvious. And these differences have increased over time. The contrast between twenty-first-century humans and other animals seems undeniable once you're in the business of sending a sports car into space.

Much of this is driven by culture. Culture's achievement is to store information outside the body. Ants have reached their global success by diversifying into more than fourteen thousand species, but humans have few environmentally specific adaptations. *Homo sapiens* have speciated their cultures. A population of more than seven billion of us sharing in cultural knowledge is what has allowed our generation to visit the Moon, while the men and women who painted the caves of Lascaux hadn't even figured out the wheel. They weren't less smart, and our brains aren't bigger or more powerful now. It's just that we're inheriting knowledge that has been improved on over time. The other side to this, as evolutionary biologist Joseph Henrich has pointed out, is that without access to each other or culture, we can find ourselves less able to survive on Earth than many other animals would.

John Berger wrote that animals were once 'with man at the centre of his world'. Not any more. With each century that passes, people have settled themselves further away from the rest of the living world. Yes, this is not yet true of all human societies – there are still tribes in the smoking jungle of the Amazonas, groups of herdsmen on the Mongolian plains, small societies that live cheek by jowl with nature. But the fact remains that more than half of the world's population now lives in urban places. This is likely to rise to three quarters in the coming decades. The people of urban societies now spend hours absorbing news about the human world. Consuming a diet of YouTube clips of human techno-logical brilliance, we feel more certain than ever of some absolute difference between us and other animals. Yet, despite these ever more outlandish achievements, we never quite manage to remove the obstruction that the rest of the Earth creates.

As the centuries have passed, no evidence from the natural world has arrived to support our view of ourselves. By and large, quite the opposite has happened. As we've peered deeper into the enormity of what surrounds us, the notion that humans are special and separate has become more complicated. Think of the once common idea that humans are at the centre of the universe. When medieval astronomer Tycho Brahe identified a supernova, the creeping suspicion that human life must be understood in a far more unsettling context was unavoidable. In a conversation I had a few years ago with Mary Barsony of the SETI Institute, she spoke of how people react to the work of cosmologists. 'We're wired to wonder what the meaning is,' she said. 'But it's hard to push meaning onto the natural world.

Whatever animal you have, it's the centre of the universe. But we're not only constrained by what we know,' she added, 'but by what we can bear to know.' The naked truth is that we've existed in our current form for a few hundred thousand years in a universe around thirteen billion years old. Some still cling to the possibility that all the events that led to the arrival of *Homo sapiens* form part of an anthropic principle. In other words, the universe came about somehow so that we could come about. But this kind of reassurance is a gamble. A little time and exploration have tended to shatter such hopes. This is especially true when there are so many unknowns.

Chris Impey, an astronomer at University of Arizona, points to the fact that biology seems to emerge out of extreme niches. But what happens once you have biology? Are multicellular life forms inevitable under certain conditions? Or is it difficult to get from an amoeba to a horsetail fern? The relatively new science of exoplanets – planets that could host life – involves a range of assumptions about what makes somewhere hospitable. The hunt for Earthlike planets got started almost accidentally when NASA's Kepler space telescope team detected a rocky world in 1995. But the first true exoplanet, Gliese 581c, was discovered in 2007 by Michel Mayor and Didier Queloz. Since then, we've found thousands of them. On this basis, researchers have predicted that there may be ten to twenty billion habitable worlds in the Milky Way alone. These would be planets that are neither too hot nor too cold, with sources of water and energy. NASA is now searching for signatures of oxygen, water vapour, maybe even the burning of hydrocarbons.

But time is essential to life. It took several billion years before primitive animals evolved on Earth, and another several billion before we arrived at an animal with technologies like telescopes. Other animals on our planet have cultures and strategies for social learning, especially mammals. But mammals gained their opportunities to adapt in part down to release from predation when 90 per cent of life on Earth was disrupted or destroyed by a random meteorite. 'If we're the consequence of that opportunity to regain a niche,' says Impey, 'how common might such opportunities be?' For those who see it as a favourable expectation that we're not alone, it's frustrating that life may be uncommon in the universe. But others see it as a boon that we've failed to come face to face with E.T. The anthropic principle remains unchallenged while we're alone.

One obvious way to resolve the debate over the anthropic principle is to deny the whole lot. The foundational text of Rabbinic Judaism influenced centuries' worth of scholars and led to the widely held assumption that the Earth was around 6,000 years old. The attraction of a young Earth is that it looks like a human planet, a place whose only history is that of the dominance of our species. But by the eighteenth century, Scottish geologist James Hutton had recognised that the natural processes of erosion and sedimentation called for a new vision of time, a time of rock and stone, what he called 'deep time'. And news began to spread of strange objects and even stranger bones dug out of the ground. Each decade brought to light more evidence of early forms of humans, threatening the conviction of generations of men and women that they were the mortals of a God-made world of recent history. The conclusion ought to have been unavoidable: not only had the landscape around us slowly evolved but so, too, had we.

But there are still millions of people for whom such facts remain a controversy. Young Earth creationism is remarkably prevalent among some branches of Christianity, particularly within the evangelical movements that have gained traction through popular TV shows. Since 1982, the US-based Gallup poll has been asking Americans for their beliefs on the evolution of humankind. Even among postgraduates, more than 20 per cent in recent years continue to believe humans were created in their present state less than 10,000 years ago. It's a reminder of how fiercely people can use their minds to deny an unwanted reality.

There's nothing new in this. There are plenty of examples throughout history of the trouble that fresh knowledge can cause for old hopes. In the decades of the great dinosaur hunters of the eighteenth century, scientists and thinkers had argued over the consequences of the finds. The bickering wasn't really a matter of controversial data but rather a psychological response to the threat the evidence posed towards cherished religious and mythical explanations for human origins. But decades of careful study of the geology of Europe and the Americas ultimately made it impossible to deny the Ice Age that must once have transformed landscapes and through which large, now-extinct animals like cave bears then roamed. The Earth itself became a giant creation story in which the characters were written in fossilised bone and in artefacts preserved in pages of soil and rock. The Earth was speaking over us with greater authority.

But nothing could prepare people for what Charles Darwin would bring. For the man who watched Darwin steadily tease out his evolutionary ideas while aboard the *Beagle*, the relatively late recognition that we share aspects of anatomy and behaviour with other animals because we share a common origin in the long-distant past caused the 'acutest pain'. Darwin and Captain Robert FitzRoy were already at odds on the grounds of politics. FitzRoy was a Tory and a supporter of slavery, while Darwin was a Whig and an abolitionist. FitzRoy strongly denied the new geological theories of the age of the Earth and refused to be drawn into the scepticism that was beginning to affect those around him. A famous anecdote has him holding up a

Bible at the 1860 Wilberforce–Huxley evolution debate in the Oxford University Museum in order to denounce Darwin. According to onlookers, he told the crowd that 'I believe this is the truth, and had I known then what I know now, I would not have taken him aboard the *Beagle*'.

The anxiety after Darwinism was enormous. Darwin himself was susceptible to strong feelings of unease about his theories. In 1860, he wrote to the great American botanist Asa Gray. His sense that there is 'too much misery in the world' was causing him to question his faith. 'I cannot persuade myself,' he went on, 'that . . . God would have designedly created the *Ichneumonidae* with the express intention of their feeding within the living bodies of caterpillars.' If we and everything else belong to the landscape of parasites and meteors and blind exchanges of energy, then God must have made the world so. But, if that was the case, the world offered a much more complicated measure of our worth.

Much was made of the danger Darwinism posed to the literal truth of the dominant religions and creation myths of the world. But Darwin knew that his theories posed an even greater threat to moral beliefs. If we are animals, then should we see morality as natural? Given how unpleasant nature can be, how can being animal offer us any guidance in how we should act? These difficulties continue to vex us. Ever since we've understood that human life shares common ancestry with the rest of life on Earth, we have been trying to contain the spread of its significance into our moral lives. But it spreads unseen, like water damage.

Religion offers a moral source that lies beyond the entanglement of nature. But without God, our convictions look much shakier. American philosopher Mortimer Adler has expressed the panic people experience when asked to consider relinquishing the unique moral status of humans. 'Why, then, should not groups of superior men be able to justify their enslavement, exploitation, or even genocide of inferior human groups, on factual and moral grounds akin to those we now rely on to justify our treatment of the animals we harness as beasts of burden, that we butcher for food and clothing, or that we destroy as disease-bearing pests or as dangerous predators?' It hardly needs pointing out that throughout much of the history of modern civilisation, groups of self-appointed superior men did just this. Moral behaviour and norms have always been profitably fluid. Yet now it's an embarrassment to our confidence that morality exists in and is derived from nature. At all costs – the logic goes – we mustn't allow the moral nature of humans to be an animal thing.

For thousands of years perhaps the answer was to see all of life as marvellous: the vault of the grasshopper, the shivering tail-tip of a hunting cat, the coloratura of a songbird. For those worldviews loosely clumped together under the term 'animism', there's less of a gap to justify. And if a kind of magic quickens

the sinews of living things, then humans simply possess a share
in a sacred cosmos. There are still traces of this original numi-
nous landscape. It's there in a handful of hunter-gatherer cultures and faintly in polytheistic worldviews. But as a rule, it
doesn't seem to have survived the transition into large agricultural societies.

History shows us that knowledge doesn't necessarily lead to
enlightenment but only to the search for a different source of
light. Each scientific or intellectual threat to our singular status
has been followed by a fracturing of existing beliefs and renewed
efforts to ground the basis of our separation from the rest of
life. One solution was to redirect the emphasis on to becoming
human. Solace could be found in the possibility that, as Thomas
Huxley put it, we may be 'from them' but we are not 'of them'.
This has been repeated in different forms ever since. It's
common to hear that while science tells us that we are animals
subject to the same laws as other organisms, humans are
'uniquely unique'. In this way, scientists swallow Darwinism but
remain immune to its effects.

The idea is that we were not human and then we became
human. And when we became fully human, we could no longer
be understood as animals. This idea gained popularity in the
decades after the publication of *On the Origin of Species*. As
evidence of our early ancestors was found, thinkers and scientists started to focus their energies on defining the moment
when we became human as we understand it today. The desire
was for unique biological characteristics that could be dated
after the emergence of our species.

By the early twentieth century it was widely considered among
scholars that around 40,000 years ago, in an era referred to as
the Upper Palaeolithic, a cognitive leap occurred whereby
groups of *Homo sapiens* in Western Europe began acting and

behaving in a way that marked them out as human. The 'Human Revolution', as it became known, was talked about as an almost miraculous span of time in which a suite of skills such as abstract thinking, sophisticated tool use, language and symbolic image-making arose among men and women, transforming people in an extraordinarily short timescale from an ape to a superbeing. If God hadn't brought forth humans in a state of completeness, at least evolution had very nearly done so. But this source of reassurance soon hit difficulties.

If it was biological proofs of difference that mattered, how could we separate out biology from cultural behaviour? In an age of smartphones, it is obvious to most of us now that cultural innovations can accumulate suddenly without any physical change to the people inventing them. It's perfectly plausible that there was no human revolution, only some phases of rapid evolution and many phases of slow evolution. It's also possible that the underlying cognitive abilities that gave rise to the cultural manifestations we've since labelled as 'human modernity' were present tens of thousands of years before the speciation of *Homo sapiens*, let alone the arrival of the Neolithic era. As American archaeologist Sally McBrearty has since said: 'The search for revolutions in Western thought has been, in part, a search for the soul, for the inventive spark that distinguishes humans from the rest of the animal kingdom.'

Alfred Wallace, who arrived at his own theories of evolution at more or less the same time as Darwin, understood the advantages of this for human psychological wellbeing. To revive our hopes for salvation, we could explain away the body as a natural event, but single out some essence that is the source of 'a higher intelligence'. It is because of notions like this that modern secular individuals don't need a unique soul. It is enough to believe that our elaborate cognition sets the boundary. The boundary in this sense is not between an immortal being and its physical body but consists of the superior qualities of human rationality. Human mental life consists of a range of capabilities that lift us out of nature.

This was particularly appealing to those who had inherited humanist ideas, sometimes of deeply compassionate intent, to place the interests of human persons at the centre of judgement. According to secular humanism, perhaps other animals are sentient, even conscious by some measures, but they lack a sense of self, any knowledge of right or wrong. They lack a soulful mind. Experiences like pleasure were to be given intrinsic value, so that we could point to the duties that arise from this. In practice, all this did was to isolate human things and then use this to argue we only have duties to humans. It was a neat trick. Humanists had carved our statue and hidden the chisels with which it was made.

While Enlightenment humanism did much to argue for science as the true expression of human power, science in its pursuit of minutiae has refused to toe the line. The evidence from science has continued to tell us that there's no such thing as a human in this sense. The traits and appearances that define animals come about through processes. They're neither an end point nor a scale. Most of the capacities we prize evolved gradually and would have been at least partially present in ancestors that today we would consider as without any special status whatsoever.

Ian Tattersall, a veteran taxonomist and curator emeritus at the American Museum of Natural History, is acutely aware that 'biology doesn't permit neat boundaries'. Yet history has proven many of us to be 'reluctant to admit diversity'. It is more than possible that biological architecture can come before the behaviour kicks in. Take language, for instance. For humanists, language is often held up as one aspect of the unique essence that makes us more than animal. Yet there's great disagreement about the origins of language. Some of this comes down to how we define language. Among language specialists, Derek Bickerton sees true language as something that only emerged in *Homo sapiens* and as a 'catastrophic event'. Tim Crow argues it was a speciation event, and Richard Klein that it may have come even later. But American linguist Ray Jackendoff is among a group of scholars that believe language developed incrementally, beginning around two million years ago at the onset of the *Homo* evolutionary branch.

For a long time, it was also presumed that other hominin species like Neanderthals died out because they didn't possess skills like human language. Yet in 1983 a hyoid bone was discovered among Neanderthal remains. The hyoid bone is a funny little horseshoe-shaped bit of our anatomy believed to be essential for complex speech. Some have argued that the bone might have come in handy for singing rather than speaking. But others believe the evidence points to speech. If possessing language is that which justifies our special status, then we must at least acknowledge it now looks likely that this wasn't a *Homo sapiens* thing but a hominin thing. This is much more confusing. It may place the evolution of language back to a common ancestor. Recent analysis on dental specimens suggests that the two species diverged at least 800,000 years ago. Although gene flow continued between the two species, it muddies the waters if we hope for a pure source of exception. If we were to travel back in time to observe the first of our ancestors chatting together around a fire, we might see a bunch of hairy, heavily browed animals.

And there would probably be more humanlike animals than historians have cared to admit. In 2019, a complete hominin cranium was recovered from Woranso-Mille in Ethiopia by African anthropologist Yohannes Haile-Selassie. It caused upset because, up to this point, it had been assumed that modern humans evolved in a direct line from *Australopithecus anamensis* and then *Australopithecus afarensis*. Yet the specimen revealed the likelihood that these two intelligent, upright primates overlapped with each other as separate species. Human evolution appears to be highly branched, not a straight arrow of descent. Haile-Selassie and others see this as evidence that our ancestors belonged to more general evolutionary trends to adapt to changes in climate and shortages of food. The same year, Russell Ciochon's team successfully dated skeletal remains of *Homo erectus* in Ngandong in Java to around 108–117,000 years ago. A human ancestor once thought to be a direct ancestor now looks to have overlapped with our own species too.

Nobody knows the full story yet, but the idea of revolutions softens a reality that is strange and disturbing to us. If language, as argued by someone like prehistorian Robert Bednarik, evolved gradually throughout the Pleistocene period, when did we suddenly cross some unbreachable line between us and other animals? Most of what we can see in the fossil record points to the slow stages that have led to everything we esteem in ourselves. The controversial jasperite cobble is a small piece of rock that looks like a human face. What makes it of interest is that it was found with the bones of an *Australopithecus africanus* individual in a cave in South Africa. It isn't evidence for art, and we can't prove it was deliberately in this animal's possession. But how else did it get there? Jasperite isn't found anywhere near this region. What if it was noticed by a being that wasn't in the *Homo* branch of primates, and what if it was an object that meant something to this creature: a treasure?

This piece of jasperite is millions of years old. But by at least 800,000 years ago, it looks like hominins were discriminating between ordinary items and more exciting ones, like crystals. At this time, there are also possible signs of the symbolic use of pigment. Another compelling but uncertain object is the Tan Tan figurine from around half a million years ago. This seems to exploit visual ambiguity. A semi-weathered bit of a stone, it looks like a voluptuous woman. The shape is natural, but those who have studied it believe there's evidence that the grooves that give it a human form were artificially exaggerated.

The least controversial of such early indications of a more assertive consciousness comes from the creation of cupules around the world in the Lower Paleolithic. These are depressions in a rock surface, as if a small bowl has been set into it. They are made deliberately by percussion, using a hard object. Some specimens in the Kalahari Desert date from more than 400,000 years ago. Some may be even earlier. They are widely regarded by specialists in rock art as among the first efforts made by animals to express themselves symbolically. Often there are hundreds of them grouped together, like a close-up of the skin of a strawberry. A single cupule might require thousands of blows to make on hard rock. Pounding on stone takes time and energy. Why on earth did these beings do it?

Art historian Ellen Dissanayake believes these ritual marks stimulated the opioids in the brain that produce feelings of trust and security among small groups of individuals. Did the hammering sound like thunder or the hooves of a stampede? Did our ancestors sing along with the rhythm? Nobody knows. But these creatures would not only have been powerful predators themselves; they would also have been prey. Although they were not modern humans, whatever kind of mind they had was supple enough to begin ritual.

These glimmers of a complex truth matter. They matter because they show us that we are part of a gradual metamorphic act of life. The search for revolutions or for natural traits that belong exclusively to *Homo sapiens* is certainly of interest. But it is also a compulsion among those who need a solution to Darwinism. Gradualism makes morality less absolute, weakening the confident basis of our exclusive moral status. That the generations after Darwin hunted for signal markers so assiduously only exposes a deeper psychological basis to the search.

None of this is to say that we shouldn't answer to the specific capacities of humans as we are today. In whatever way it was that we evolved, humans are remarkable and it seems right that we should respond to our particular needs. But why doesn't it follow for the needs of other species too? How often do we dismiss the culture or language in other species as diverse as scrub jays and bottlenose dolphins? The work of evolutionary biologists like Andrew Whiten has revealed the extent to which other animals use social learning to mitigate nutritional stress. Chimpanzees, who have been known to use over thirty different kinds of tools, exploit natural hammer materials when their fruit diet is depleted in the dry season. Orangutans use a look-and-learn method between mothers and their children to pass along important survival tools, like using stems for getting at termites. Chimpanzees also fish for termites and have been seen donating tools to teach less able youths in their group, at a cost to themselves.

This is worth bearing in mind given that more than 60 per cent of primates are endangered because of our behaviour. Since the 1960s, populations of chimpanzees have dropped by a half. In the end, we do little to halt these losses because we believe in an absolute border between us and them. Their deaths are but a candle snuffed out. We forget that the recent ancestors to whom we owe our life would be dismissed by the same measures today.

On being a moral predator

Hundreds of years of denying that we're an animal has left us with inconsistencies and confusion. Yet efforts to redress the balance have often led to vociferous denials that we can make

sense of our lives in biological terms. When entomologist
E. O. Wilson suggested intellectual fields could be synthesised
into something he called sociobiology to begin the proper
scientific study of human nature, a public intellectual spat in
the broadsheets gave rise to a bitter debate. Some of the
objections to the claims of sociobiology were based on
contested views of evolution. Such pushback was motivated
by concerns about too simplistic a view of the plasticity of
animal behaviour, humans included. But some of the strongest
reactions flowed from an instinctive aversion to the idea of
humans as animals. Students demonstrated against professors.
Individuals who supported the idea of the biological study of
human nature were subject to aggressive jibes. In one lecture
in the late 1970s, an anti-racist group poured a jug of water
on Wilson's head.

Some of this can be explained by the proximity of the
suggestions to the horrors of the Second World War. Not only
had science been twisted by fascist ideologies to justify racist
fantasies, but a good deal of genetic science had flourished
under the distressingly permissive research environments of
Nazi Germany. Anything that sought to explain or define people
through their biology strayed into this region of horror in living
memory. What people had come to dread was the ranking of
human life based on biology.

Despite this, much of the modern world is still supported
by hierarchical myths of one kind or another. Most religious
societies have viewed life as a chain of being, with God at the
top and men tucked under the angels. We've never shaken free
of this ladder of life. Secular humanism promised to liberate
us from the delusions of faith. But humanism has erased God
and drawn the self-actualised individual in a heaven of his own
making. In the 1930s, the German émigré neurologist Kurt

Goldstein argued in his book *The Organism* that animals are driven by the motive to realise their full potential. Humanist psychologists like Abraham Maslow visualised this as a pyramid in which people rise from the fulfilment of base animal needs to the final achievement of their full cognitive capacities.

After Darwin, this sort of hierarchical thinking ought to have been left behind. Yet it has persisted. We've largely ignored Darwin's scribbled note to himself not to use the words 'higher' or 'lower' in discussions. This is especially true of mental processes, which we routinely label as 'higher order' or 'lower order'. Many of us now assume that only one kind of thinking has intrinsic value. Or, more drastically, that there's only one kind of thinking, and only humans can do it. This is not only wrong; it impoverishes the living world.

Every day we gather more evidence for the obvious marks of feeling and intention in the lives of other creatures. That other animals don't think and behave like us or have the same needs as we do is significant only up to a point. What we can see is that animals contribute to the overall success of the living environments of our planet and that they have vivid, complex lives that involve feelings and mental control. Even insects get a look-in. A study published in 2016 in the *Proceedings of the National Academy of Sciences* by Andrew Barron used scans of the brains of animals like honeybees to show that the midbrain pulls on memory and perception to support behavioural choice. The usual retort is that human consciousness is nonetheless more complex and therefore has higher value. But ranking biological traits is hardly in our best interests. Humans may have a weakness for hierarchies, but they rarely serve us well. Even a cursory glance at history makes something plain: while humans have tended to have an idea of humanity as special, they have not always agreed on who gets to be human.

Today, it is commonly written that the human brain is the most complex object in the universe. We owe this oft-repeated quote to James D. Watson, co-discoverer of DNA, and one of the great biologists of the twentieth century. For both Wilson and Watson, humans are animals that can be understood through due attention to the fact that we are biological beings. But where Wilson hoped a recognition of biology would enlighten us, Watson believed it could enhance us if we wished. For Watson, science, especially genetic science, can make a better world of healthier, smarter, more beautiful people. In the late twentieth century, he argued that parents should be free to exploit technologies. He threatened litigation against the *Telegraph* newspaper for misquoting his belief that women should have the choice to abort if they believe their child's life will be ruined by being homosexual. But in the Channel 4 documentary *DNA*, he argued for the benefits of selecting out the 'lower 10 per cent who really have difficulty, even in elementary school'. People, he said, believe these poor achievements are down to poverty but 'it probably isn't. So I'd like to get rid of that, to help.'

In all this, it is useful to know that Watson was one of the signatories of the third Humanist Manifesto, written in 2003. The first manifesto was published in 1933. At the time, the movement was called 'religious humanism', to foreground humans as the creators of their own existence. Religious humanism was striving towards 'the complete realization of human personality'. Institutions were there for 'the enhancement of human life'. All thirty-four signatories were prominent men.

The Second World War put rather a dampener on this confident assertion of humanity's natural eminence, but in the 1970s a second manifesto appeared. The word 'religious' had been dropped. Acknowledging the gross misuse of industrial technologies that had yielded the concentration camps, this was nevertheless the 'dawn of a new age, ready to move farther into space and perhaps inhabit other planets'. Technology can 'control our environment . . . unlock vast new powers'. Again, the goal was the growth of 'each human personality', through which we might give 'personal meaning and significance to human life'.

The mistake made by the humanist movement was in refusing to let go of the fundamental dream of human transcendence. Humanists needed something inborn, something non-religious but still exceptional to keep hold of a hierarchical vision. But in seeking to make humans special by rational introspection, they discovered only the ways a human animal relates to itself and others. There is only really one stage of our lives that contrasts dramatically with other animals. And so, slowly and fitfully, humanist thought constructed a world in which this stage is paramount, the phase of social consciousness that gives us the notion of a person. Personhood became a twist on the soul for those who still wanted their own version of heaven. In doing so, they landed humanism with some knotty problems. What of those humans without clear attributes of personhood?

Most of the things we define as unique aspects of our mental life are lacking in babies, corpses, or those in comas, all of whom we now see as warranting respectful care and treatment. They are also lacking in some of our most severely disabled people. From one angle, once we accept that we're animals, things get a little easier. When we acknowledge that humans have a life cycle, we can make better sense of our instinct to care for the severely disabled or what should ground the basis for our actions towards embryos. Different stages of this cycle make separate demands on us. It's only when we single out traits that we lose sight of the truth that, as animals, we are dynamic and various in our forms.

When we accept that humans are animals without resorting to claims of superiority, we can appreciate and value those of us that genetic inheritance or mutation have affected, from conjoined twins to people with Angelman syndrome. Disability or ageing are not seen as threatening aberrations from the essential existence of a human but as normal occurrences among organisms, which is not to say that we do nothing to ease associated suffering or to better their lives, but that we don't see these individuals as diverging from a superior definition of being human. Nobody is expected to conform to some median of what an adult human might be or to be measured

against the maximal state of human capacities. But this wasn't the path taken by humanism. Instead, secular forms of salvation sought to pick a difference in nature and turn it into a justification. But if there is something in our biology that makes us the most or even the only important animal, how do we stop the use of biological traits as the basis of how we treat one another?

It was into this chasm that the Nazis entered. Aktion T4 (the name came from the abbreviation of a street address where plans were drawn up) was an enforced euthanasia programme that ran from at least 1939 to the end of the Second World War. The first T4 site was at Brandenburg, where thousands of children were killed because, in Hitler's phrase, they were *lebensunwertes Leben*, 'life unworthy of life'. This wasn't new. A treatise, *Permitting the Destruction of Unworthy Life*, was published in 1920 by Karl Binding and psychiatrist Alfred Hoche. And such ideas weren't restricted to the rising National Socialists of Germany. In 1927 in America, the Supreme Court ruled that there was nothing unconstitutional about the compulsory sterilisation of mentally or physically 'retarded' humans. In an article on this history of persecution, American writer Kenny Fries said of his own disabilities, 'Those of us who live with disabilities are at the forefront of a larger discussion of what constitutes a valued life . . . Too often [our lives] are not valued.' Not only are they undervalued, Fries tells us, they are feared. Disabled bodies become a disavowal of a fixed template of human superiority.

Since the Second World War, plenty of people have considered dignity as a powerful intuition that supports the protection of all human life. As a concept, it has become a catch-all for the kind of thinking we hope to engender in individuals and social institutions. But dignity as it's imagined at present is only a mental contrivance. It has surfaced as a means of justifying why those without evidence of subjective consciousness can still matter to us.

Take the Universal Declaration of Human Rights. It stipulates that everyone has the right to recognition as a person before the law and that we are born equal in our dignity. In his *Oration on the Dignity of Man*, the Renaissance philosopher Pico della

Mirandola has his God declare to humankind that 'we have set thee at the world's centre'. Dignity restores humans to their central position in a world shaken by knowledge. Dignity may be an important intuition for how we ought to treat one another, but in its current incarnation it is another property that only humans can possess. The more promising possibility that what we call dignity might be a way of readying the mind to be respectful of others, human or otherwise, has been ignored by those who don't want to be an animal.

James Watson believed that his luckiest break was having atheist parents who had 'no hang-ups about souls'. By the time we arrive at the pithier version of the Humanist Manifesto drafted in 2003, we are told that humanism is a progressive philosophy, relying on science as the apotheosis of progress, and with ethics deriving from the 'inherent worth and dignity' of human life. Today, secular humanism is the largest alternative to religious dominion, but it is another myth of authority. For non-believers, a range of often absurd or partial explanations have had to be concocted to justify imposing hierarchy on the Earth's vibrant confederacy.

On the one hand, humanists declare that we are the consequence of evolutionary processes along with all other animals. On the other hand, their faith that our lives and deaths matter more than those of other animals has nothing to do with evolution. Their assumption that we can use other animals as we wish has little foundation in science. While humanist thought is careful to ensure that we don't rely on measures of superiority among humans, the flourishing of other animals doesn't matter, at least not at an individual level. This is because we have decided that our natural features are more significant than theirs.

Humanism has relied on the notion of intrinsic values,

the idea that what appears unique to our species can form the basis of inalienable moral principles. We are forever arguing that animals lack this or that capacity and that this lack can ground some or all of our reasons for how we behave. We look for lines in the sand. The ability to suffer. The ability to have intentions. But we fudge the grey areas that threaten our paradigms. We either ignore the threat or try to come up with some special conditions that make sense of why, in our case, we cannot base everything on any one of these abilities. There are endless, high-flown arguments about this, some of them fiendishly clever. But all of them struggle to survive if they are based on the idea that a part of us isn't animal.

The true source of much of our moral confusion comes from something surprisingly simple. Many of the tensions we experience derive from the dissonance inherent in being a predator with a rich moral faculty. That we belong to those animals that seek and kill other animals for food and yet are capable of amazing tenderness and emotional depth is an important part of our natural history. Killing and the seeds of kindness abound in the wild, most especially among mammals. Foxes crush beetles between their sharp carnassial teeth without a flicker of remorse, but the vixens that aren't breeding that year will look after their sisters' offspring, play with them, feed them. The young foxes, lifting their faces to the sunlight as they loll in the grass, undoubtedly know what it is to feel pleasure in the world. How different might it have been if our kind of intelligence had spun out of herbivores? What if a moral system had emerged among the beavers in Canada? Or in populations of manatees swimming silently through the Amazon or – as they now do – at the edge of Florida's nuclear power plants?

It is a testament to the depth of our denial that we are animals that we still can't admit this. All that we do, we do as animals. But we justify it as humans. The way we've structured the world is little more than the intuition of an animal whose greatest interests lie with its own kind. But we almost never confess to this. We tell ourselves instead that we have a soul. Or, if we don't want to call it soul, we say there is a person inside us that is more important than the body from which it's made. In this way, the bodies of other animals mean nothing. They are bits of machinery, spare parts. We may use them as we wish.

Our entire value system is a cave of our own making. In English, the modern usage of the word 'animal' as a creature separate from civilised humans emerged at around the same time as humanism. As Alasdair MacIntyre has written: 'From its earliest sixteenth-century uses in English and other European languages, "animal" and whatever expressions correspond to it have been employed both to name a class whose members include spiders, bees, chimpanzees, dolphins *and* humans . . . and also to name the class consisting only of non-human animals. It is this latter use that became dominant in modern Western cultures and with it a habit of mind.'

It's worthwhile to note that no matter how deep into the physical substance of organisms we go, we never find an edge. There's no absolute dividing line between us and other animals, nothing that can offer us an unassailable moral category. Science now shows us not only that the lives of other organisms are far more complex and sentient than we've sought to admit, but also that our own traits – to lesser or greater degrees – are things that nature can reinvent. A blow to God, perhaps. A far greater threat to a rational animal convinced it's the centre of things.

While the slightest insight into our experience shows how necessary it is to care about human lives, this doesn't separate us from the rest of our planet's life. Human life remains an animal life. Our awareness and our exchanges of ideas make us a powerful and sensitive animal, but none of this brings our animal life to an end. That this idea has had such extraordinary influence across our societies is only now becoming clear. It is surfacing because it matters more than it ever has that ten thousand years of modernity have delivered an animal that doesn't think it's an animal.

Primroseworld

In April 2019, an Israeli spacecraft crash-landed on the Moon, potentially depositing tardigrades from our planet. Tardigrades are tiny, 1.5mm extremophiles – creatures that thrive in harsh habitats. There was no public consultation for the contamination of the Moon with life from Earth, nor was there any well-reasoned ethical underpinning for the act. What existed was the money and technology to do it. The tardigrades, captured for the voyage on the back of sticky tape, can survive in a state of suspended animation until conditions are suitable for survival. The team behind this event was the Arch Mission Foundation, whose aims are to maintain a 'backup of planet Earth' in order to 'disseminate humanity's most important knowledge across time and space'. The designers of this experiment hope that no matter what happens to human civilisations, their archive, engineered as a durable record of our presence, will be packed off around the solar system. It will 'guarantee that our species and civilisation will never be lost'.

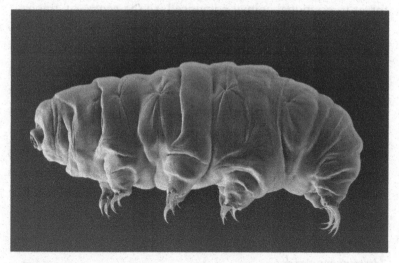

There are a few similar projects: the Long Now library, a supersized memory of civilisation in the form of an archive, the Memory of Mankind, and the EarthLight Foundation, which aims to carry 'the seeds of life to places now dead'. The EarthLight Foundation is a loose collective of mostly wealthy individuals with a mission to 'support the expansion of life and humanity beyond the Earth'. EarthLight claims to 'treasure and celebrate life'. Meanwhile, the Arch Mission Foundation was founded by Nova Spivack, a technologist and futurist. His reasoning? 'Help us save humanity.'

Elsewhere we have the Svalbard Global Seed Vault buried, with more than a little irony, in an abandoned coal mine deep in the permafrost of Arctic Norway. Inside are a wide variety of plant seeds stored in case of a global crisis. As well as plant life, there are gene banks in which the sperm and eggs of animals are contained using methods like liquid nitrogen, in the event that we should need to restore biodiversity. Biorepositories like this began to crop up in the aftermath of the Cold War. They've taken on a new significance as people have become more aware of the effects of human actions on the rest of Earth's life. Faced with a landscape of extinctions, our minds have inevitably turned to the potential of our own.

In 1970, on the first Earth Day to mobilise action to prevent the dwindling abundance of other species, an astonishing twenty million Americans took to their streets to demonstrate. I wasn't born then and there has been nothing else like it since. Meanwhile, in my lifetime, the already small numbers of some of the most iconic species, from lions and tigers to giraffes and gorillas, have continued to shrink. Tigers are now at around 7 per cent of their historical numbers. It is much the same for the lions in Africa. In the summer of 2019 the world woke up to the publication of a new collaboration between Chinese and Australian scientists that found that more than 40 per cent of the Earth's insect species could go extinct this century.

Humans have always affected the landscapes and natural systems around us, but there is a big gap between the effects of a tribe in the Serengeti and many millions of us with machines. To give some sense of the changes that come from developing more powerful technologies – changes which are not, of course, only about losses – we can look at epiphytes. Epiphytes, amazing but often overlooked organisms in our world such as mosses, liverworts and orchids, are plants that grow on other plants. In my home country, where industrialisation began, evidence suggests that populations of epiphytes were 80 per cent higher before the transition to new manufacturing processes in the late eighteenth century.

Some have called this epoch of disappearances the Anthropocene, Paul Crutzen's definition of the new age of humankind. It is perhaps best for now to stick to the Holocene extinctions. Spanning at least from the death of the woolly mammoth to the vanishing of the passenger pigeon, the ways we are causing these shifts in the populations of living organisms are complex and various. But what we can say with confidence is that we are having a disproportionately large effect on the Earth and most of its organisms.

Our struggles with being an animal make it difficult to appreciate our role in changing life on our planet. They make it even harder to figure out what we should do. When it comes to the declines in species, some voices have pointed out that extinctions are natural events in the evolution of life. Some have even argued that the current phase of extinction may ultimately have its merits in creating the conditions for a new phase of diversity. Trilobites, which dominated life on Earth for two hundred million years, met their end in an epic extinction event in which more than 90 per cent of marine animals died out. It didn't happen overnight, more like over tens of thousands of years. But whatever the ultimate cause – too much carbon dioxide in the seas, a sequence of massive volcanic eruptions, another asteroid – it allowed for the rebound that saw massive, terrifying reptiles and the emergence of warm-bloodedness in some early mammals. In turn, mammals thrived after the extinction of nearly all dinosaurs.

As the dominant animal, we've had a role in reducing the diversity of mammals again. The recent studies of Matt Davis, Søren Faurby and Jens-Christian Svenning have estimated that around three hundred species have gone extinct as modern humans have spread into new regions. All of these species were part of ancient lineages. In calculating how many years of evolutionary history were lost after these extinctions, Davis and the others arrived at a figure of two and a half billion years. Should we feel bad about it? But we are part of nature, some say, so our effects on the environment are natural. Let nature take its course.

This line of thinking is assisted by the fact that our actions are beneficial for some species and not for others. Sometimes we rid the world of orchids and mammals, but other times we flood it with primroses. *Primula kewensis* is a hybrid of one plant that grows on the limestone cliffs of Yemen and another from the Himalayas. It has chalky, toothed leaves and pretty flower heads the colour of ripe lemons. It emerged like a rite of spring at the turn of the twentieth century from the world's largest garden. Kew Gardens, dug into the old grounds of royal estates in London, is a concoction of grandiosity and paradise summoned in the decade before Darwin had disordered the minds of Victorians. The Kew primrose is evidence for evolution at work, a new species that has arisen accidentally because of us, a pretty outcome of our actions.

People cling to sunny stories like *Primula kewensis* as a reminder that to mourn the disturbances and losses of other species is to have a rigid ideal of nature. They argue for the persistent, non-discriminatory movements of evolution, by which new life forms and half-breeds come about. The point here is that the effects of human activities on living things aren't uniform. No matter what we do, some species might benefit, but others might die out. The rationale is that life is a mercurial force, not a permanent state. Those who argue from such logic believe that we are being selective and arbitrary in how we see other species and landscapes.

But those who make such arguments don't apply the same logic to us. It's easy to see why. It includes the idea that it would be equally natural for us to go extinct. When it comes to human lives, we don't like the argument that something is good because it's natural. If it derives from being an organism or is part of the earthly or cosmic processes in which a beetle or a mountain have shares, we quickly realise that this confuses the worth of many of the things we aim to keep in check. Nature is a crooked yardstick against which to measure choice. For us, something being natural can only be a selective justification.

Nowhere is this starker than in today's conservation movement. In 2018, the New Zealand government began the Predator Free 2050 policy to protect the native biodiversity by eradicating stoats, rats and possums, which arrived on the islands with humans. Despite concerns from scientists that some of the underlying assumptions were flawed, the policy has larger problems in its logic. For one thing, what exactly makes biodiversity a good value when the individual lives of animals appear to have little or no moral weight? New Zealand, after the dinosaurs died out, has been a landscape of endemic species, largely birds and insects, with few mammals. Its flora has also been particular to the islands, with its trees among the largest in the world. The kauri tree has the greatest timber volume of any of the Earth's species. Unsurprisingly, when the human mammal, or more specifically European settlers, populated the islands, almost all the kauri were cut down.

Polynesian rats came with the first arrival of Maori, causing the extinction of many animals, especially ground-nesting birds. Maori also hunted animals in large numbers, including the iconic moa bird. European colonists brought with them domesticated cats. More extinctions followed. Overall, the arrival of humans into New Zealand led to catastrophic losses of species and ecosystems. Now the emphasis is on predators unwittingly brought into the country. Less is said of the enormous impact of the country's dairy herd or the agricultural pollution of the rivers and lakes by humans. The value of biodiversity is still discussed in vague terms concerning the benefits of a healthy, balanced ecosystem, whether because other animals and land-scapes assist in flood control or carbon storage or pollination. But these so-called natural services make less sense of why we might shoot a stoat to save a nightjar. In what sense does the nightjar interest us? Nobody yet answers that adequately.

In any case, life is a difficult starting point. The EarthLight team have declared the importance of treasuring and celebrating life. However, is it only multicellular animals – or are we including life forms like *Shigella dysenteriae*, a form of bacteria that causes a nasty dose of stomach flu? It feels easy enough to value the oak tree in our back garden. It's more difficult to rustle up much love for the powderpost beetle that makes holes in our oak furniture.

Humans in flux

But there is a greater reckoning on the horizon. Both religious worldviews and the various forms of humanism are riven with anxieties about our own bodies. In each of these worldviews, there's rough agreement that our bodies, so nakedly animal, aren't what matter about us. There's something extra that is the true living substance. And this extra element keeps us safe from all that is disturbing about being an animal.

For those that see the body as a holy vessel in whom the soul temporarily lives, the body in all its forms, from a conceived cell to a corpse, has a special value. For humanists, our minds possess skills that no other animals have, which give us dignity, our moral faculty, our gift for reason. Science and education, as supreme forms of rationality, are then viewed as the prime means for refining humans. In between these two views are bitterly contested grounds.

These struggles are plain to see in the fusion of biology and engineering. In 2016, Klaus Schwab, founder of the World Economic Forum, published his treatise on the Fourth Industrial

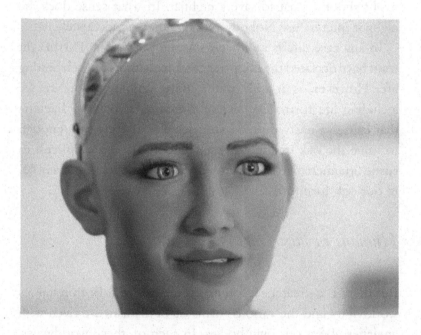

Revolution. This next phase of progress includes genomics, artificial intelligence, synthetic biology, robotics, and engineering at micro- or nano-scales. We are told that these new industries have the potential to lift all of humanity into a new phase of

collective happiness and wellbeing. At the same time there is growing concern that, somewhere in all this, we might 'lose our hearts and souls'.

As the technological gulf between us and animals widens, we feel more reassured than ever that we are unique. Yet, simultaneously, scientific knowledge proceeds to close the gap between us and other species. In this revolution, we are no different from the other organisms in our laboratories. It is our physical animal inheritance that we now rely on to make our human lives better. In this new era, we throw ourselves at the mercy of natural processes we see – consciously or otherwise – as hostile.

In his book *Visions of the Future*, American economist Robert Heilbroner argued that human societies have had three distinct phases of conceiving of the future. For a long time, we saw it as essentially static. Then, spurred on by scientific advances and a heady dose of confidence, we began to see the future as a wonderful boon. Now, spooked perhaps by the world wars and the aggressive tempo of change, we conjure visions of lost control, of possible peril, of uncertainty or ambivalence. Some of the most common spectres of our imagination concern the human form distorted in some way, deformed or harmed by uncontrollable energies, whether by plagues or our own inventions. These fears haven't come out of nowhere. Members of our societies are broadcasting their feverish anxiety about the new technologies in equal measure to those who deliver euphoric messages of unprecedented opportunity. It's a warning sign that something is out of whack. It's no coincidence that it is now that the perplexing relationship we have with being animal is coming into focus.

The engineering of biology creates problems for any worldview that denies, in some way, that we are animals. In 2003, scientists and entrepreneurs successfully sequenced the human genome, ahead of schedule. It was an extraordinary international effort to apply to humans what had been glimpsed first in 1911 by Alfred Sturtevant while mapping genetic linkages in the fruit fly, *Drosophila melanogaster*. It was hailed as the ultimate breakthrough in medical science and as a turning point in human history.

Then, in 2005, French scientist Alexander Bolotin published his work on the bacterium *Streptococcus thermophilus*, which has a novel enzyme called Cas9 that enables the microorganism to recognise and deal with invading viruses. This discovery formed the basis of a new technology, CRISPR/ Cas9, which exploits this enzyme as 'molecular scissors' that cut DNA in a particular part of the genome, allowing for the insertion of a change that will be incorporated into the genome as the damaged DNA repairs itself. It heralded the invention of a cheap and reasonably easy method of editing the genomes of life. By altering a DNA sequence, be it a plant or an animal, genome editing might be used to prevent the inheritance of a disease, to make a new material or even a new animal.

Prior to this, in the 1970s, the idea of synthetic biology had begun to take root. This is the unification of biology and engineering such that novel gene arrangements can be created in the lab. It raised the prospect of what synthetic biologist Niki Windbichler describes as 'rational design of biological systems for purposes'. In 2013, the J. Craig Venter lab, a private commercial enterprise, claimed to have made the first entirely synthetic bacterial cell. This entailed the insertion of a synthetic genome into a pre-existing cell. Yes, it was incredibly small, but this was classified as the first deliberate creation of a new life form. And so, in our laboratories, we now have the means to sculpt evolution – from the gene to the genome, the stem cell to the single-cell organism – to improve human lives.

In 2018, the research efforts in genomics reached another landmark. It was announced that a team of Chinese scientists had successfully cloned the first primates. Zhong Zhong and Hua Hua, two long-tailed macaques, were created by taking the nuclei of cells from another macaque and inserting them into an egg with its own nucleus removed. The resulting embryos were then implanted into an adult macaque acting as a surrogate. The two baby macaques were the only survivors out of more than sixty other pregnancies.

The idea behind these experiments is to utilise cloned monkeys in biomedical research to enable humans, in the long run, to live better and longer lives. Some people marvelled at the cuteness of Zhong Zhong and Hua Hua. Others recoiled in horror. We have a mixed history with the long-tailed macaque, native to the rainforests of southern Asia, sometimes denigrating them as pests and sometimes commending them as sacred. But their role as lab animals is prized because they are primates like us. It is this fact that underpins the horror experienced by some people. If we can clone macaques, we can clone humans.

From one perspective, a cloned human would be no different from any other human, in as much as the individual would be a person made from human biological stuff. But if the world is based on some essence that forever separates us from other animals, the idea of a clone is feverishly provocative. In Kazuo Ishiguro's novel *Never Let Me Go*, a clone describes how just looking at one of them was enough to give an ordinary person 'the creeps'. In Ishiguro's imagination – and in the minds of many of those who consider them – clones elicit a visceral response, a sense of repugnance.

The biomedical work of human cloning at present has nothing to do with raising the embryo to the stage of a walking, talking adult. But the fact that we could, in theory, is enough to reveal the unstable territory we've created for ourselves by founding everything on our difference from the rest of life. Little wonder that the United Nations Declaration on Human Cloning soon followed. It began as an attempt to organise a world ban, including therapeutic cloning, but settled as a non-binding declaration. According to the declaration, cloning is 'incompatible with human dignity and the protection of human life'.

We saw the beginnings of this anxiety with Dolly, a sheep

that became the first mammal cloned from an adult somatic cell. Shortly after Dolly's birth was announced, there was a flurry of public debate about cloning and those at the frontline of the technologies began to receive letters and calls from women hoping to act as surrogates for their dead parents and other such solutions to the intolerable transience of human life. Ryuzo Yanagimachi, who began cloning mice in the early part of the new millennium, was targeted by those at the frontline of despair. 'A couple whose young boy was dying after a car accident asked me to take a cell and clone him,' he told reporters.

In 2002, Pope John Paul II issued an edict arguing that cloning reduces a person to 'a mere object'. 'When all moral criteria are removed,' he said, 'scientific research involving the sources of life becomes a denial of the being and the dignity of the person.' Around the same time, the Raëlian cult in Canada publicised their belief that all humans had been bioengineered by a superior alien civilisation. The group mounted a concerted effort to develop the resources to undertake the private cloning of a human, in some kind of epiphanic moment for humankind. They managed to involve a young couple who had lost their baby in a case of hospital negligence. It didn't come to pass.

Over twenty years on, the initial panic around cloning has softened enough for research to progress. But uncomfortable hypocrisies remain and continue to trip us up. Nowadays, we have little legislation to prevent the cloning of pets to assuage the muddled grief of their owners. Companies like ViaGen Pets in Texas provide their clients with clones of much-loved dogs, cats and assorted others for eye-watering sums of money. The practice is possible because other animals lie outside of our jurisdiction. We must fulfil certain agreed limitations on suffering, but the inevitable deaths that follow in a messy process like cloning are seen as acceptable collateral.

It is reasonable to assume that our sensibilities will seek other loopholes in a permissive commercial environment. Ten years or so after stem-cell technologies became widely known, illegal clinics began to proliferate, offering lucrative treatments that had not undergone clinical trials. It must be expected that in the coming years something similar will take place with other technologies. These, too, will likely be driven by a combination of greed and grief.

In December 2019, Chinese scientist He Jiankui was fined and jailed for his role in creating the first gene-edited babies by modifying heritable parts of their DNA to be resistant to HIV. The response around the world to his experiments was fast and brutal, and he was found by Chinese authorities to have pursued 'personal fame and fortune, with self-raised funds and deliberate evasion of supervision'. Yet He's work took place in a wider context of heavily invested breakthroughs in China, including the first gene edits to non-viable human embryos in 2015.

Any survey of industrial history ought to counsel us that we are rarely good at predicting risk. What we feared would happen often never comes to pass, and what really bites is what we never saw coming. But if enough of us don't think that what matters about being human has much to do with being animal, we face a future push to exploit technologies that will control or eliminate any biology that stands in the way of what we want. And if enough of us continue to think about natural traits and propensities in hierarchical terms, the future pressure for human enhancement will become difficult to withstand.

Of course, it's a savvy move to present biotechnology as being a wonderful opportunity for those that have been traditionally marginalised, whether they be HIV-positive, disabled or female. But it's most likely that it will be the few and the powerful that will profit. In 2014, Facebook and Apple announced egg freezing as a benefit for female employees. Google followed. A full-page cover of *Bloomberg Businessweek* shouted: FREEZE YOUR EGGS, FREE YOUR CAREER. It was marketed as the ultimate hammer to the glass ceiling. But it's an easy get-out for the harder work of meeting the different health and financial needs of women.

Thinking like skin

In stripping us down to our fundamental materials in order to find some way of repairing or enhancing the human organism, we continue to expose to light the fictions on which we've based our experiments. Sarah Franklin, who has written extensively on the confused ethics surrounding biotechnologies, notes that 'the difficulty of defining life leads to the use of several, often conflicting, and inconsistent, ethical models as guides for moral conduct'.

The ways that humans have legislated the ownership of their own life came from the liberal democratic tradition of humanists like John Locke. The principle of self-ownership paved the way for many social changes we prize, including the emancipation of women and the abolition of slavery. But who can be said to own the body parts for which there's no legal person present? In the case of *Moore v. Regents of the University of California*, a disagreement erupted over the use of body tissue derived from John Moore's spleen cells to produce a new pharmaceutical application. The case resulted in the prohibition of individual ownership of one's own bodily tissue.

Moore brought the suit against his doctor, David Golde, the researchers and institutions that used his cells, and the Sandoz pharmaceuticals corporation. The court dismissed the idea that a person has unqualified rights to the products of their body because the cells are 'no more unique to Moore than . . . the chemical formula of hemoglobin'. They also determined that the property was not the cells, but the cell line later created from them. However, much of the decision rested on the fears that any extension in property rights would stymie a large amount of medical research.

The fact is that a human skin cell or spleen cell is both living and yet intuitively of little value until a technology arrives that can make use of its hidden vigour. We didn't evolve to have insight into or to develop moral relationships with cells, but with human individuals. And we have based our legal systems on the idea that it is personhood that matters. This is entirely understandable given the important role that what we call 'a person' has in how we communicate our needs to one another. But the emergence of person-like traits is a natural event in the whole life cycle of a human. It isn't a tidy line.

The social moral consciousness of humans emerges when it's needed. It is in the functioning adults of our species at reproductive age that social negotiations and transmission of social learning are essential. It doesn't mean that a toddler has less value than we do, whether or not they will ever grow into a healthy adult. As Franklin writes: 'The use of biologically based timelines to establish ethical parameters . . . does not "resolve" the underlying ethical questions. It is undeniably the case that a two-week-old human embryo is a being, and that it is a form of life. That it is a living human being is therefore incontrovertible. Yet it is no more or less a living human being than an egg or sperm cell, or, for that matter, a blood cell.' And if a skin cell now has the potential to be turned by us into a person, how should we think about these almost invisible traces of life's potency?

Consider the new technology of in vitro gametogenesis. Most

of us are familiar with IVF. This method of manipulating human reproduction has been integrated into societies as a norm despite initial fears. But many will not yet be familiar with the idea of IVG. After embryologists first observed fertilisation in the late nineteenth century, the orthodox opinion was that only an egg fertilised by a sperm could lead to a live birth among humans. But efforts to understand the exact mechanisms of fertilisation have resulted in a new technology on skin cells that scientists hope can remove the barrier of eggs for women with limited reserves who can't have children. In 2017 Japanese scientists used IVG to make baby mice from their mother's skin. This led to headlines about men having babies with each other and women having babies without men. What these new technologies make clear is how malleable is the matter of which we're made.

In the case of humans, IVG promises, in theory, a future of clinics with the technology to manufacture an almost limitless supply of sperm, eggs and embryos. In practice, this could allow for the selection of embryos from a supply of hundreds rather than tens of embryos. In turn, this could 'raise the specter of "embryo farming" on a scale currently unimagined, which might exacerbate concerns about the devaluation of human life', wrote the authors of the 2017 paper.

Those who believe in the soul resist such flagrant interference with God's work. Those who believe in the humanist notion of dignity urge much of it on because it might improve the lot of a human person. The harm done to any other species along the way is largely a matter of paperwork. Ethics committees are springing up in universities around the world tasked with making sense of how we categorise the entities we are creating in laboratories through techniques from stem-cell biology to genome editing. Legal precedent and policy tend to exploit ethical uncertainty or hesitation. Our technological breakthroughs are supported by the assumption that humans have full moral status. We mustn't kill or harm each other for the sake of better healthcare. This is often argued from our cognitive prowess: our ability to be aware and rational, for instance. But how do we make sense of chimeras with brains that are partly composed of human cells?

Currently, we license scientists to insert human cells into some animals, like pigs and monkeys, in the hope we might harvest organs from them, yet we are unable to clarify exactly why we should ensure there can be no emergence of human traits, cognitive or otherwise, into these transgenic creations. The UK Academy of Medical Sciences 2011 report 'Animals Containing Human Material' was confident that substantial modification of an animal's brain to make it more 'human-like', or experiments that could lead to the animal looking like one of us, should be guarded against.

In September 2018, overturning earlier policy, the American National Institutes of Health confirmed it would not support studies involving 'human–animal chimeras' until after a review of the scientific and social implications. In its statement, the agency foregrounded the worry that animals' 'cognitive state' could alter under the presence of human brain cells. This followed the discovery that scientists had already begun experiments with the assistance of funders like California's state stem-cell agency. Human stem cells had been injected into animal embryos and gestated in female livestock.

The first ever transgenic mouse was created in 1974 by inserting a DNA virus into a mouse embryo to show that the manipulated gene was present in each cell. Since then transgenic mice, rats, even goats and pigs have been modified for a host of different purposes. Perhaps nothing 'human-like' will emerge in these

genetic hybrids that might present us with further moral difficulties. Yet some of the research using human–mouse brain
chimeras has shown that the hybrid mice can both learn and
recall with greater aptitude than those without human cells. How
might any new moral measures affect both our chimeras and us?

One of the first historians to see the uncertainty is Sophia
Roosth. She spent many months researching among geneticists
and synthetic biologists and is articulate about the confusion
that has arisen. 'A question I heard voiced by scientists in the
lab,' Roosth says, 'is, "What kind of entity is this? What does it
mean to have these new forms of life that contain genes from
really diverse lineages?"' As philosopher Michael Sandel notes,
new technology 'has induced a kind of moral vertigo . . . we
need to confront questions largely lost from view in the modern
world – questions about the moral status of nature, and about
the proper stance of human beings towards the given world'.

The dilemmas we face in what we're doing to the living
world, including our own bodies, aren't about economics or
resources or data. They're echoes from a world we'd prefer to
ignore. We are only just beginning to learn of the difficulties
these new powers create for an animal that has spent much of
recorded history denying it is one. At present, we still assume
that human interests lie outside of nature. Yet it's common to
hear that it is to nature we should default if we find ourselves
feeling squeamish about new technologies. Given that we
already share nearly half our genes with a banana, why should
we protest against the insertion of some kind of amphibian
gene that might give us an immune-system boost?

What this new era of technologies has revealed is that there's
nothing essential about the stuff from which we're made. Animals
are more like cell colonies than pure genetic vessels. And life
seems to be a gloriously fluid and permutable thing. It's a quixotic

process of readiness. It is opportunistic, potent. Nudge it ever so slightly, and extraordinary things can happen. A few years ago, veteran geneticist George Church caused a furore when he made a playful comment that an 'adventurous female human' could, in theory, act as surrogate for a cloned Neanderthal baby. To soften the comment, he later remarked that an alternative scenario could be the modification of a chimpanzee such that one of the female primates could carry a Neanderthal hybrid to full term. In all this, Church is seen as a rabble-rouser. Yet Church is only expressing an unpalatable truth that is still denied more than a century after the death of Darwin.

Church's half-jest about Neanderthal surrogacy – aside from the unspoken fact that it would rely on human genes to bring back another species – was mirrored in another jousting thought experiment by David Barash. Barash, an American psychology professor, put forward the idea that we might use new technologies to create a humanzee, a human–chimp hybrid. It made international news because Barash argued that the experiment would teach 'human beings their true nature'. And so we begin to ask whether humans are as impressionable as clay, made of a chimerical stuff that can take on the genes of another animal or be rewritten in the shape of desire.

The essence of something seems to evaporate when everything from sex to species can be manipulated into a transformation given the right biochemical prod. Technologies like cloning and IVG worry us because we slip into the moral category of other animals. We're no longer miraculous individuals but a bit of matter that can be selected, altered and used for unforeseen purposes. It's a bold reminder that we're 'of life' as much as we're from it. As an animal, what we are is in constant flux. In this gift for metamorphosis there is no essence, no mark of justice. There are only profound contradictions.

THE CIVIL WAR OF THE MIND

There existed a natural man; an artificial man was introduced within this man; and within this cavern a civil war breaks out which lasts for life.

Denis Diderot

On persons

It has been assumed for centuries that what makes humans special is our gift for thought. Yet we neither know nor agree on what 'thought' is, or 'intelligence' or 'consciousness'. Still, today, most of the uses of our biological traits in both law and ethics originate in assumptions about the value of our special mental capacities. This is often summed up by the idea of being a 'person'.

In many legal definitions, a 'person' can be natural or artificial in as much as they are a fellow human to whom basic rights are owed and who, in turn, has certain duties; or an entity, such as a corporation, that has duties and must abide by the law by dint of being an artefact of human society. In philosophical terms, a person is often conceptualised by the kind of subjective consciousness humans experience. What is often meant by this is that our memories and awareness, our sense of self and autonomy, are the essence of who we are.

But most people probably think of a person simply as another human being. Some perhaps extend something like personhood to their pets or to other animals that seem to share a high level of consciousness. And, in some cultures, ideas about what we are and what experience means can be radically different. Still, regardless of time or culture, it nonetheless feels as if we think because there's a 'me' to do the thinking. And this can give us the impression that what we ultimately are is the thinking bit of us. In other words, mind maketh man. It is this 'person' who possesses the mental skills that are often held up as the pinnacles of experience – reason, intelligence, morality. It is a person who can participate in the laws of societies, own property, or be the citizen of a state. Out of this extend the threads of the structures and systems we have strung across our planet.

But what follows is that because what seems to matter about us is our state of personhood and this, we are told, is a mental experience, the rest of our animal experience or physical reality can be considered as less important, even as separate to who we truly are.

To Aristotle, the human mind seemed 'a widely different kind of soul, differing as what is eternal from what is perishable'. In a heartbeat, Aristotle gave a compelling reason for why we wander about in clothes discussing metaphysics while other animals still do their business in the dust. From those early questions about the striking differences in human and animal lives, the soul migrated into the mind. The soul inside all living things became instinct.

Still, as the philosopher Helen Steward told me, that's not entirely fair to Aristotle. 'It may be true that Aristotle over-egged the rational aspect of the soul, but . . . the psyche wasn't a *thing*.' For Aristotle, if we are animals, then persons are animals, not some kind of abstraction inside us. But this subtlety has largely been lost. By the Middle Ages, a person was defined as 'individua substantia', an individual substance of a rational nature. This idea had emerged in Europe as part of an extensive theological debate about the nature of God. By 1689 Locke defined a person as 'a thinking intelligent being, that has reason and reflection, and can consider itself the same thinking thing, in different times and places'.

This line of thought suggested that our thoughts and memories were a kind of soul-like thing that might be detached from the body. And it offered a patch of dry land for humanity above the confusing, unscrupulous waters of nature. If what we shared with other animals couldn't define us, then human morality and ethics might have a modicum of absoluteness. And, if we could establish that all other animals lacked what goes on in our minds, indeed lacked mind altogether, then we would be alone not only as moral agents but also as moral subjects.

Today, we talk confidently of the unassailable value of human persons. The soul in the skull is a mark of justice. In its absence, we matter less. By its presence, we become capable of self-determination: a subject, a citizen. We are lifted out of the bondage of nature and into the liberty of reason. But, in an era of biotechnology and artificial intelligence, there are several things worth mulling. Over thousands of years of modern civilisation we have attempted to rank and classify the whirling, indefinite psychological intelligence that illuminates us in a self-aware, perceptual storm. But what the actual properties of a person might be remain uncertain. Does a person need to have language? Do they need to be smart or moral? Do they need to be human or have a body? The truth is that what we mean by a person shifts over time and context. This isn't to say that our sense of a thinking self isn't real but only that it's not like an element that can be extracted and weighed. Yet, it is this will-o'-the-wisp that gives us a legal and moral standing that no other animal has.

For some, then, being a person is what makes humans unique. For others, anything can be a person, since they say it's a physical illusion that a sufficiently intelligent machine might replicate. The first group believe the soul in the skull makes us human. The second believe humans can make a soul. The only point of agreement is that the body of an animal has little or no significance. It shouldn't surprise us if such a view puts us and other forms of life in jeopardy.

Mental give and take

In our more enlightened moments, we recognise that our ideas about persons and personhood aren't quite right. If the important quality of personhood is having a sense of self – the old idea that there is something 'it is like to be' – what happens when this is present? What, in other words, does being a person do for an animal? Somewhere along the way, we have forgotten to ask why biological life forms might be such things as persons. Or why they might experience the sensation that they are one. One thing's for sure, being a human person isn't only about our own mind and psychology. It seems to be as much about knowing others as it is about knowing ourselves. It's more like an action than a thing.

To keep ourselves alive, evolutionary history has allowed us a personal glimpse of our own bodies and experiences. In turn, we have the scope to see into the experiences of others. Our minds seem to be mind-generators. We conjure a rich, history-infused knowledge of our own motives and needs and experiences, but also an insight into others. And so, what we name as persons are not only the knowledge that we ourselves have a mind, thoughts and experiences; they are the insight that others do as well. A person, in this second sense, is formed of our perception of others as having thoughts and feelings. For humans, it takes one to know one. This insight into ourselves and others gives us much that we have come to call moral. When we speak of being generous, what we seem to be doing is generating another's lived experience. And it doesn't just have to be humans' experience. We can imagine the experiences and intentions of other living things and even, in our dreams, of machines.

But there are good reasons to believe that the human experience of being a person is an open-ended and flexible affair. Our kind of intelligence has emerged through the realities of being an animal entangled in its environment. When we give a name to the self-reflecting part of ourselves, we personalise a range of processes sheathed in our animal bodies. Inside the shadow of identity is a mix of episodic memories, images and hormones. The insight that someone or something else has a mind or thoughts or feelings seems to be an ancient way of enabling an otherwise moderately aggressive primate to get along with others. But what seems to really define us is that humans have social relationships that can swiftly change.

Humans are borne along on invisible currents of conflict and companionship. Like some other mammalian predators, such as wolves and some bears, we are extremely social. Indeed, by the standards of most social predators, humans are extraordinarily cooperative. We live in large groups of related and unrelated individuals, and we work together to an incredible degree for a species of primate. But sociality is often met by counter tendencies or mechanisms. There are limits on our willingness to see into the lives of others, most especially if we want to use or attack them. As such, humans have a stock-pile of mechanisms that both bring us together and allow us to disregard one another. The human social mind is a supple affair, skilled at switching allegiances as much as making them. There's something restive in human psychology, a mental bargaining that is thorny and tangled.

There are whispers of this in the breathtaking diversity of ways in which we behave towards each other. In most individ-uals, the private narrative of our selfhoods is capable of aggrandising, loathing, pity, shame, ambition and contradiction. Yet the fickle motives are glossed in social mores and insignia that somehow keep all this chaos together. Through shared symbolic worlds, we achieve intense cooperation with one another. We undertake tasks both intimate and immensely complex. We can help a total stranger whose car has broken down and we can also work together to build a bridge across a vast body of water. But think of the digital mycelium of the internet. It ought to have been a testament to teamwork. But what is taking place is an unedifying display of rudeness. What might have been proof of the human propensity for conviviality and exchange has turned into an algorithmic tribalism.

Every hour countless millions of us cooperate and communicate with one another. Nonetheless, a study by psychologist Robert Feldman found that 60 per cent of people lie at least once in a ten-minute conversation, and on average two or three times. Both men and women lie frequently, but woman are 'more likely to lie to make the person they were talking to feel good, while men lied most often to make themselves look better'. In a separate study, Charles Honts, a polygraph expert, determined that people lie in about 25 per cent of their social interactions. Some of this is flattery and social finessing, but the point here is that human relationships are often defined by guile. Evolutionary biologist Robert Trivers and psychologist Dan Ariely have both studied human deception, separately concluding that people not only lie to others, but they lie to themselves the better to lie to others. The logic goes that if you can convince yourself, you're more likely to convince someone else.

This is to say nothing of the damage we can do each other. One of the greatest injustices of the modern age is to objectify someone. We owe many of our assumptions about objectification to the humanist philosophers of the Enlightenment. Following Kant, American philosopher Martha Nussbaum isolated several key features of objectification: an emphasis on someone as a means to another's ends and a denial of humanness or personhood. Researchers have repeatedly found that those who are objectified are more likely to be depersonalised. In other words, the sense that another of us has meaningful thoughts or feelings is much weaker when they are someone we want to use.

Slight changes in power structures can upset people's relationships with surprising speed. In the 1970s David Kipnis intuited that power-holders devalue those considered lower down in a hierarchical situation. In his experiment, some participants were assigned as supervisors of a specific task. Half were told they had extra powers to use over those beneath them; half weren't. Kipnis's study found that those primed with the idea of their greater power were almost always influenced by it at the expense of those they supervised. Around the same time, psychologist Philip Zimbardo conducted a simulated prison experiment. He found that his college students started treating their prisoners in repressive and dictatorial ways if they were given the role of a guard. This tendency to depersonalise in conditions of power can become magnified when people are tools in a political fight. Think of the practice of necklacing in the apartheid struggles in South Africa. Or the waterboarding of insurgent Filipinos by US forces at the turn of the twentieth century. No society is immune to such flare-ups of brutality.

But we rarely admit that such things have much to do with the alliances of a primate. We say instead that social and historical conditions have orchestrated actions. Our political leaders and legal systems make no mention of the role being animal may play. It is easy to disrupt this confidence. Consider how morality might work if we had a different kind of life cycle. What might property rights look like if we were eusocial insects? This is a term most often taken to mean an intense level of social organisation, commonly found in highly related insect colonies where a single female or caste produces offspring. How might personhood function if we were colonial animals like siphonophores, which seem like one creature while being made up of thousands? Subjective experience, our sense of duty to one another, would be quite different for creatures that are a congregation of semi-autonomous parts. If we accept that these kinds of natural histories would affect our value systems, we can admit that ours are founded in part on our biology.

When we define a person as a rational individual who exists across time, a self-determining agent with rights and duties, we squeeze people into a shape for which they weren't made to fit. Human persons – whatever exactly they are – have been summoned by forces that seek options, not reasons. We generate a person in the mind to empathise with others. We distort the person in the mind to thwart empathy. And sometimes we deny there are minds, despite all the evidence to the contrary.

From a me-mind to a we-mind

The origins of our social psychology are not at all certain. The real lives of animals are so embedded in a finely threaded chaos that it is almost impossible to unpick any single cause from another. But if we walk back over two million years to southern and eastern Africa, we come upon a land growing a little cooler, less densely wooded. Our ancestors, along with other species living alongside us, needed new strategies to survive the changes. The first walking primates had a brain volume not dissimilar to that of chimpanzees today. This had doubled by the time we arrive at *Homo erectus*. The brains of most modern humans are around three times larger than those of chimps, although smaller than those once possessed by Neanderthals.

Anthropologist Stephen Oppenheimer has signposted Sarah Elton's work, which shook 'our sense of uniqueness in human life'. She measured the brain size in a number of fossil skulls from primate species over a million-year period from around two and half million years ago. She drew on species from two main branches of hominids – *Homo* and *Paranthropus*. She also looked at *Theropithecus* monkeys, who had a diet of fruits, nuts and tubers. While the brain size of the monkeys stayed stable, both branches of hominids showed a dramatic phase of brain growth. This is particularly interesting given the fact that *Paranthropus* apes are most likely to have been vegetarian. If we aren't looking at brain growth in only one direct line of succession, we must assume that something, or several things, were causing mental intelligence to be continually selected for.

One theory is that some species of primates faced intense pressure from predators in the Pleistocene and therefore more variable conditions in which to find food. To overcome these stresses, these animals began to coordinate their behaviour in larger groups. In turn, this led to a reliance on social learning, the handing-down of useful behaviour in a lifetime rather than through bodily instinct and habit. This supports what we know of chimpanzees' tool use and social learning: that it increases when a key food source is depleted.

Of course, nobody exactly knows what happened in our prehistory. But what is certain is that the *Paranthropus* line died out. It's a dizzying idea to wonder where their big-brained path might have taken them had they survived. Today only we remain as the latest and solitary *Homo* species. We can only guess that having a sense of self and judging the states of mind in others helped our ancestors to know, learn, reflect and manage behaviour when group living had become a matter of survival. But we make a mistake when we think that human persons are the consequence of cooperation. It's true that humans are social animals who survive by living in groups. But this picture shouldn't be understood too simplistically.

As a collection of individuals, we're not a superorganism. Humans are only ever conditionally groupish, and our mental life comes into clearer focus when we understand that we don't always want to cooperate. This is true not only for individuals, but also more broadly for the group. Groups of people help each other out, but only up to a point. They also compete. Indeed, it's not unreasonable to consider that we are the only remaining human of several different species because we not only out-competed but actively killed our relatives. As palaeontologist Nick Longrich has said: 'We wonder what it might be like to meet other intelligent species, like us . . . It's profoundly sad to think that we once did.'

The kind of self-awareness we have, that ordinarily develops as we age, has undoubtedly refined as a consequence of being a group animal. It is not only a deeply biological phenomenon, but the result of an animal negotiating and re-forming relationships. Among humans, our perception of others is modulated by the kind of relationship we wish to have. It's less cooperation per se and more cooperative versatility that defines us. This makes a difference. Our personal point of view assists us in exploring our own motives and feelings and in making inferences about those of others. But once our minds took this step, darker moves became possible.

The word 'coalition' is medieval French in origin, meaning 'the growing together of parts'. It comes from an earlier idea of fellowship and the Latin term for uniting. Behavioural ecologists study a widespread phenomenon among animals that is referred to in the literature as 'coalitions'. These are groups of mixed individuals who cooperate for a common goal or against a common enemy. This kind of complex behaviour is typical of highly social species like dolphins, chimpanzees and us.

There's a huge amount of research on how and why animals, like lions, form coalitions. More recently attention has turned to the temporary alliances between humans. Like other animals, we protect ourselves by cooperating with those we see as part of our alliance and fighting those outside it. Coalitions, particularly among men, and rivalries between different groups are a ubiquitous feature of human societies. American psychologists John Tooby and Leda Cosmides have argued that much that we emphasise as unique human behaviour – our shared sense of culture, our insights into ourselves and others, our intense systems of moral behaviour – would all be useful in the kind of social ecology for which coalitions are suited. Coalitions, Tooby and Cosmides suggest, can manifest in a wide variety of forms including ethnic and religious conflict, castes, gang rivalries, social clubs, team sports, video games and, most acutely, politics and war. As Tooby has said, it's a 'rich, multi-component' reality. But for this to work, there must be mechanisms that enable temporary cooperation. This is particularly so when a community of related and unrelated individuals live together, full of different tendencies, motives and a background of constant change.

At present, it's controversial how coalition strategies are adaptive, but the different camps who study coalitions nevertheless agree that they are. The theories on morality developed by Jonathan Haidt and Jesse Graham are based on the belief that primitive moral intuitions among early ancestors coordinating in groups were the origins of later moral beliefs and values. For Haidt and Graham, moral intuitions provided an adaptive advantage to our ancestors by supporting social order for group living. By following a set of innate 'moral' rules, groups of animals reduce levels of conflict and raise levels of cooperation. Another adaptation might be the extending of the idea of an agent to refer to groups and communities.

Other researchers place the emphasis on how a relatively small primate might compete with dangerous predators such as lions and hyenas. Michael Tomasello has theorised a way in which the minds of animals began to unite, sharing their intentions for specific tasks and finite periods of time. He points to the collaborative behaviour of animals such as pilot whales, killer whales, dolphins, Harris's hawks, wolves and meerkats. These animals must have some knowledge of themselves and the intentions of others in their group. But in humans, this kind of cooperative mind-sharing may have led to collective intentionality, from a 'me' mind to a 'you and me' mind to a 'we' mind. By this theory, group life became a shared activity, creating, in turn, a shared worldview. Or, some have suggested, a shared mind for the group, almost a shared sense of a person.

But strategies and social behaviour are unstable, swithering as conditions shift. The possible sources of conflict happen at all levels of the individual and the group, between relatives and strangers, between men and women, and so on. We can see evidence for the separate motives and tactics among humans in the fact that we are sexually dimorphic. Men are proportionally stronger than women, on average 75 per cent stronger when it comes to upper-body strength. In cross-cultural surveys on physical aggression, men are considerably more likely to use this as a tactic.

Consider the crucial matter of when to have sex. If everybody's interests and efforts were aligned, this would be a perfunctory matter. But that's not what we find. Gray langurs are a slender species of monkey living across much of India, with a shock of white hair around a small black face that give it the look of a wise elder. Like humans, langur females conceal when they are fertile, unlike many other mammals that show off their fertility by exposing genitals, by the swelling and redness of sex organs or the release of pheromones. Most species with concealed ovulation have some degree of promiscuity, as well as the possibility that the males may kill babies that are unrelated to them. Most also have babies that require round-the-clock care, which means both males and females are essential to the investment necessary for raising the child.

Primatologist Sarah Blaffer Hrdy has done much to increase our understanding of the cooperation and conflicts needed to resolve the immense pressures females face in raising young that are vulnerable for so many weeks or, in the case of humans, years. At the same time, Pascal Gagneux has found that more than 50 per cent of the offspring born to chimpanzees in his study were fathered by males from other troops. The females had sneaked away around the time of their oestrus, only to turn up again a few days later. Alert to their fertility, female chimpanzees attempt to evade the notice of males in their troops and access alternative sexual choices. As cognitive scientist Denise Cummins wrote of the findings, 'This observation perhaps offers the clearest testament to the impact greater intelligence can have on reproductive success.'

Among humans, then, male and female coalitions will be different. One of the most obvious causes of shifts in power among individuals is whether assets must be shared or can be monopolised. One kind of asset is those with whom we wish to reproduce. In early hunter-gatherer societies, the emphasis on coordination between males was probably greater. But in early complex societies, like those of the Aztec and Incan civilisations, higher-status men enjoyed greater access to women. When resources must be shared, egalitarianism becomes almost inevitable. But if resources can be monopolised, tolerance and insight might be forsaken in the sudden scramble for power.

For British biological anthropologist Richard Wrangham, the kind of group living that human ancestors evolved after splitting from chimpanzees likely involved something akin to self-domestication. Work on domestic phenotypes suggests that some cooperative behaviour needn't be selected for but will emerge as a byproduct of selection on systems of emotion or aggression. Both Darwin and Franz Boas had already noted that humans share domestic phenotypic traits. Some more recent and intriguing work by Simon Kirby and James Thomas has shown that domestication in the Bengalese finch, foxes and dogs leads, over time, to increases in gesture and vocalisations, as well as lower aggression.

The neural underpinnings of aggression show remarkable cross-species similarities. But it's a mistake, Wrangham says, to think there's only one kind of aggression. Aggression can, for example, be either reactive or proactive. Reactive aggression comes in response to a direct threat; proactive aggression aims to control or coerce, to gain dominance or status. Chimpanzees appear to have both high proactive and reactive aggression. But humans have low reactive aggression and high proactive aggression, something affected by cortical regulation. These are signs that something radically changed in the way humans interacted with one another to succeed. Wrangham theorises that our male ancestors killed those who were toxic to group cohesion. But some of these changes may also have resulted from the preference for choosing partners, both sexual and cooperative. Psychologist Alicia Melis found that matching temperaments in chimps altered the outcome of tasks, so that cooperative tasks were more successful when high-tolerance chimps were paired together.

Whatever the exact processes by which humans came about, it seems likely that once our ancestors began cooperating at a certain level, they became their own environment of selection. This was the conclusion arrived at in the 1970s by biologist Richard Alexander. 'The human brain has evolved in the context of social cooperation and competition,' he wrote in his seminal paper 'The Evolution of Social Behaviour'. Inside cooperative groups, deception is key. Alliances can be made in a clandestine way with other individuals through food sharing or grooming, favours that can be called in another time. Assessments and renegotiations constantly take place in the relationships between all members of a group. For this, it's useful to know something of what someone else is thinking or scheming. And it's also necessary to have some measure of self-control.

Alexander claimed that the reason social competition continued to exert an evolutionary effect is because the risks of leaving the group were severe. But this didn't mean it was all about benefit. Early humans likely endured a number of shifts between cooperative and competitive social hierarchies. What's more, associations across the borders of groups will have required an even more volatile psychological terrain. In 1976 English philosopher Nick Humphrey argued in his paper 'The Social Function of Intellect' that the human brain and our 'runaway intellect' evolved to deal with the complexities of social interaction. It seems likely that our kind of consciousness may have been filtered through time in a kind of arms race. What may have benefited the individual may not have tallied perfectly with the needs of others in the group. Through social learning and tool use, parenting and group coordination, humans began to reduce Darwin's hostile forces of nature. But this opened the way for other humans to become the principal hostile force. Instead of adapting in response to a hostile environment, we became our own landscape to out-think.

When we recognise that we are social primates whose beliefs about the world and others change to fit the relationship we want, our hypocrisies become more predictable. Our social psychology not only determines our relationship to other living things, but has profound effects on how we treat each other. The use of hierarchical ideas to justify our relationships with each other and other living things is a legacy of our social behaviours.

The snake on the cave wall

Nowadays, differences in beliefs about human uniqueness can lead to fierce divisions between people. But for the rest of life on our planet, the result is the same. Other species are of only limited worth. They exist below some vital threshold of meaning. Meanwhile, according to the most widespread views, we are special beings with souls or minds that give us a shot at immortality. Much of this turns on the idea of being a person. Qualifying is often down to assumptions about stronger or weaker forms of experience. We tell ourselves it is the prized capacities of our personhood that ultimately convince, and therefore only we have true value. Only our lives and deaths have significance. Other animals can be killed and die without dignity. They concern us only up to a point.

Regardless of whether such a view is justifiable, it's widespread among those who wish to use other animals in ways that distress or destroy them. Meanwhile, we spend thousands on the wellbeing of our companion animals, give them names, mark their birthdays, mourn their deaths. The golden retriever that sleeps at the foot of someone's bed is no more intelligent or special than the pig they may have eaten earlier that day. But the retriever is in partnership with a human. What the dog and the pig have in common is that their worth has been arbitrarily given on the basis of the relationship a human wants with them. The dog and the man are friends, and so the dog is given identity and a mind, of sorts. The pig, just as sentient and sensitive a creature, can be rendered unconscious, hung from an overhead rail and slit behind their jowl to sever both jugular and carotid. Much of this comes down to judgements about the kinds of minds and experiences other animals have.

We know this. Popcorn movies about complex encounters with intelligent alien species entertain precisely because we can imagine these creatures as persons, even though they aren't human. In these fantasy worlds, we glimpse our judgements of another's humanlike mental attributes. In recognising other persons in this way, we seem to rely on a complex interplay of biological processes. Unsurprisingly, this is subject to powerful bias.

We call the ability to observe our own behaviour and create mental representations of it in working memory 'monitoring'. It allows us to adjust our behaviour in response to our thoughts and memories. In simple terms, we summon an image of ourselves to give us greater flexibility in how we interact with one another. But psychologists Michael Corballis and Thomas Suddendorf have noted that human memory is unreliable. It doesn't look like it evolved only as a means of recording the past. Inspired by the theories of neuroscientist Daniel Schacter, Corballis and Suddendorf suggest instead that 'its function is to build up a personal narrative, which may provide the basis for the concept of self'.

This concept of self is perpetuated by an inner voice and by the conversations we can hold with others about who we are. Put another way, we support the idea of ourselves through narrative. Meanwhile, Schacter and his colleague Brendan Gaesser show that mental imagery is critical not just to remembering the past and thinking about the future, but also to moral choice. Images merge, mapping the self across time and place. As an example, when presenting participants with a situation depicting another person's plight, the ability to imagine helping them or to recall a related past event of helping others increased the intention to come to the person's aid.

But in order to imagine helping someone, we need a window into another's mind or experiences. Sometimes called mentalising and sometimes called theory of mind, the experiments of Uta Frith and Christopher Frith have established how people think about the minds beyond their own. Someone as young as eighteen months might have some idea that other people have mental states. But by the age of five or six, children are explicit in their attribution of minds and persons to other beings. Certain areas of the brain are consistently activated: the medial prefrontal cortex, the temporal poles and the posterior superior temporal sulcus. According to Frith and Frith, these parts of the brain may have different roles in the whole complex process of how we come to see and judge another thinking individual. The superior temporal sulcus, an area that some believe interprets visual social information too, may be where the detection of agency takes place in the brain.

This research got kick-started when American psychologist David Premack and his colleague Guy Woodruff published a paper called 'Does the Chimpanzee Have a Theory of Mind?' Premack was curious about whether other primates make assumptions about states of mind in the way that is intensely important to humans. Since this paper, Frith and Frith and others have isolated the medial prefrontal cortex as 'concerned with the representation of the mental states of the self and others decoupled from reality'. Mentalising isn't just a matter of coming to know your own thoughts, feelings and beliefs; it is also about making successful guesses about the states of others.

We might assume this is only about seeing subjective minds

in other humans, but we seem to have the ability to stretch it to include any kind of intelligent presence. In this way, another 'mind' can include the embodiment of intent in just about anything. Either way, once a personal sense exists, both conscious and unconscious manipulation can create a wider array of behaviours. Memories might be stitched together or distorted to create imagined events of future social interactions. I have often speculated as to whether mental images might be deliberately used to switch on different neurochemical responses. Robert Bednarik has suggested that visual ambiguity, like seeing a snake when it's only a tree root, could be manipulated intentionally in a self-reflecting animal: 'The cognition involved is deeply rooted in mental processes found in numerous animal species, such as flight reactions to the silhouette of a bird of prey.' If so, perhaps we might visualise a human enemy as a threatening animal in order to make use of the physical changes an encounter like this could create. This kind of manipulation could take many forms.

Our bodies have a series of hormones like oxytocin that change our minds in bottom-up processes and that our minds can alter in top-down processes. And these hormones have a remarkable impact on how we behave. The work of Carsten De Dreu raises the possibility that neurobiology may have a flipside. In activating systems that help organisms distinguish friend from foe or meal from mate, we reorder our own minds and our willingness to see minds, or the feelings, needs and intentions of others. Shutting down the aspects of our bodies that increase insight is partly accomplished by 'tightening brain-to-brain synchrony' among those in the same group. In turn, this increases the production of hormones like oxytocin that play a role in bonding. We think of oxytocin as the 'love' hormone. But it has also been found to lessen the impact of automatic responses to signs of distress in those outside the group bond. In other words, oxytocin doesn't just encourage us to bond; it prompts exclusivity. For De Dreu, seeing into the internal states and wants of others can then become an indirect way of defending against those who are competitors or those we wish to use for our ends. Of course, the same hormones can still be serviced by the body later to allow empathy to flow again beyond the borders of the group. But this isn't inevitable.

If we can mess with mental images and assumptions about minds to switch on different neurochemical responses and behaviours, then we reach a possible explanation for one of our most universal and unpleasant of habits. One of the horrors we can commit is to dehumanise someone by thinking of them as an animal or as something that has fewer signs of mental or emotional content. For this, we exploit images of other animals, especially those we fear or find disgusting. In Albert Bandura's work on moral disengagement, he and his colleagues conducted experiments in which participants were asked to deliver shocks to people. Those who had been spoken about as if they were less human were given more shocks.

Kant was one of the first philosophers to pay attention to the idea of dehumanisation. For Kant, it was a matter of seeing someone as a means rather than as an individual of intrinsic worth. Few philosophers since have paid a great deal of notice to this widespread aspect of human behaviour, but philosopher David Livingstone Smith has revived discussion about the phenomenon. In his book *Less Than Human*, he quotes Morgan Godwyn, an Anglican clergyman who, in the seventeenth century, wrote of African slaves as 'Creatures Destitute of Souls, to be ranked among Brute Beasts and treated accordingly'. In its most extreme form, this is the kind of brutal dehumanisation that not only reduces empathy but actively motivates violence. Whole groups of people can be stripped of mind in such a way that, when violence beckons, they can be killed. In 1993, a broadcast on Radio Télévision Libre des Mille Collines included the phrase 'You have to kill the Tutsis, they're cockroaches.' The Rwandan genocide followed. Twenty years later, Zsolt Bayer said of the Roma in Hungary: 'These Roma are animals, and they behave like animals . . . These animals shouldn't be allowed to exist.'

For Livingstone Smith, the problem originates in the idea of essences, a fundamental, invisible property that is specific to different kinds of living things. Humans, he believes, intuitively think in essences. In his view, true dehumanisation not only denies humanity but attributes a subhuman essence to another being. This mechanism could involve the replacement of a human essence with that of a potentially dangerous organism like a rat or a virus. This may play out in the body.

Anything that manipulates oxytocin may tip us into the more relaxed and open state in which we begin to affiliate. But this can also go the other way. We may trick our bodies into doing things that our minds might otherwise encourage us not to do. Writing on left-wing terrorism in Italy, Donatella della Porta provides an account of how those with opposing views were dehumanised, seen as 'pigs' and 'tools of the capitalist system'. In an associated study on the techniques of terrorists, Angel Rabasa and colleagues demonstrate how forms of dehumanisation can provide 'ways of reducing the psychological cost of participation in terrorist organizations'. The subtleties in the way we look at someone or something else suggest that mental images and the meaning we give to them may be a crucial instrument. At low levels, the people of most societies see those in another group as having fewer signs of human uniqueness than their own group. As a rule, said American sociologist William Graham Sumner in his book *Folkways*, it is found that native peoples call only themselves 'men'. Others are something else. The word *Deutsch* translates as 'person'. The word for the Maasai people is *hadzabi* or 'I'm a person'. The word *Inuit* roughly means 'people'.

Inside a coalition, we come to see our worldview as richer in its inherent intelligence. But, following this, we can also see those with other worldviews as having less mind than we possess. In a 2019 study led by neuroscientist Pascal Molenberghs, fMRI scans, which measure brain activity by sensing changes in blood flow, showed compelling differences in how those in one group respond to the words, faces and actions of their own group members versus people from a separate group. In effect, we find it easier to think into the mental states and experiences of those who belong to our group. On the other hand, we think less about the mindset of those on the outside. In a study conducted by psychologist Jonathan Levy on Israeli and Palestinian adolescents growing up in countries with long-standing conflicts, both sides 'shut down the brain's automatic response to the pain of those on the other side'.

The experiments of Jacques-Philippe Leyens have revealed how people judge primary emotions like fear and happiness as more primitive and animal-like. In contrast, secondary emotions like guilt and shame are more uniquely human. These secondary emotions are seen as signs of a more complex mind at work. In the experiments, people see more secondary emotions in those who are in their group, whether these emotions are positive or negative. All cultures and ethnicities seem to do this. In America, black and Hispanic people have received a disproportionate amount of dehumanisation in the public discourse. But after Hurricane Katrina, black and Latinx participants in a study attributed fewer negative secondary emotions to white victims of the events. In other words, they assumed that white victims suffered less, regardless of the actual traumas experienced. This, in turn, affected their likelihood to volunteer for or help white victims.

The work of Susan Fiske and Lasana T. Harris has revealed how people use mental shortcuts that influence the extent to which they view someone as human and worthy of respectful engagement. This has been backed up by neurological imaging. In one study, people regarded drug addicts and homeless people as low in measures that Fiske developed as universal signs of humanness. When viewing these people, the medial prefrontal cortex wasn't activated. This is a key part of the brain involved in social cognition. Simultaneously, regions of the brain associated with disgust – the insula and amygdala – became active while viewing those seen as possessing few or no positive human attributes. In this way, vulnerable people can end up caught in a vicious cycle of being the most in need of assistance and care and yet viewed by others as the least worthy of it.

As a self-reflective primate, we are, in novelist Iain M. Banks's words, 'a little bit of the universe, thinking to itself'. That these thoughts are imperfect, sometimes irrational, is not a failure on the part of our logic but a condition of why we have our personal perspective in the first place. Once you have an image of the self that binds into memories of the past and imaginings of the future, minds become things that can be played with. Seeing into, diminishing or denying the experiences of others is a crucial ingredient in our behaviour towards others.

When we consider that our social psychology has primed us to both seek out and deny the minds and feelings of others, we can start to make some sense of why seeing our minds as our souls is more problematic than the original idea of the soul itself. It's not too much of a reach to imagine that, by inventing myths of human uniqueness, early societies may have unwittingly provided metaphors that underlying selfish aspects of our psychology could draw on. When being human is about having superior qualities of mind, it falls on the exploitative part of us. Uniqueness becomes a form of ammunition. In foregrounding the idea, we may have promoted a powerful weapon for prejudice.

The importance of seeing-in

In the twenty-first century, we don't know how many animals we kill, although some estimates place it at around three billion animals globally each day to feed us. The countries using the majority of animals in research are the UK, Australia, the USA and Canada, China and Japan, and many of the countries of Western Europe. Again, we don't know the true numbers, but certainly over one hundred million animals each year. We justify our uses of other animals because we say they don't possess minds that might make us regret doing so. And we assume that what we gain from using other animals negates the harms we cause because our own flourishing is paramount. We don't often broadcast that it is because the concept of human flourishing suits our purposes.

Many of the world's legal systems have instituted the need for some animals to be saved from the excesses of human cruelty. But this isn't because they have any kind of intrinsic value like personhood or dignity. Such essences, we are told, are only found in humans. And so severely disabled people have been given legal capacity as persons and supported decision-making in their best interests. Animals, we are told, have no interests.

Such a view is not inevitable. Other cultures have counted the minds and bodies of all animals as creating different but significant forms of agency. In *The Mind in the Cave*, archaeologist David Lewis-Williams imagines the moment a seventeenth-century Frenchman, Ruben de la Vialle, enters a cavern in Ariège, in the wooded shadows of the Pyrenees, and scratches his name and the 1660 date of his visit alongside what he and others assume is the graffiti of recent tourists. But it isn't graffiti. Instead, these were the artworks of ancient hunter-gatherers. At this time, Lewis-Williams reminds us, Western thinkers had no inkling of the prehistory of human life. Little wonder that de la Vialle thought nothing of vandalising an exquisite prehistoric monument. He didn't know 'the significance of what he was seeing, so, in effect, he did not "see" it at all'.

By the nineteenth century, when Don Marcelino Sanz de Sautuola discovered the Altamira cave on his Spanish estate, uniformitarianism and evolution had begun to change people's baselines. But his family's story is one of cruel failure. We like scientific discoveries that kill cancers but not those that kill cherished beliefs. People then, as much as today, can be particularly vicious when repudiating truths that they can't tolerate. When Sautuola's find of prehistoric art became known, an initial wave of interest quickly transformed into an aggressive defence against prior orthodoxy. He was laughed to his deathbed as either a hoaxer or as the victim of a hoax. And yet the people who had denied and defamed him didn't have the truth on their side. The mesmerising images of animals in Altamira, along with those in the now world-famous caves of Lascaux and Chauvet, were painted thousands of years ago by ancient humans.

To enter Altamira is to pass into a kind of Freudian chamber, a strange, suppressed region of the mind. The cave itself isn't particularly large. To reach it, you slip through a small doorway in the rock and into a hollow of stalactites that seem to reverse time with each slow drip of water. This is where some of our ancestors sat together around a fire, sheltered from storms, stitched clothing, shared meals, sang together, who knows?

The ceiling is now covered with colonies of actinomycete bacteria that shimmer when a light is passed across them, turning the dark rock into a planetarium of sparkling silver and gold galaxies. It's a magical place, and this is before entering the famous polychrome chamber. Head away from the light of the entrance and turn left down a narrow corridor and you will pass into a long room, the roof split down the middle by a large, charismatic fracture. On either side of the fracture, and often deliberately exploiting it, are stunning paintings of animals in charcoal and ochre. The sheer number of them is intoxicating.

But it's no good standing and gazing at them. It's best to crouch or, better, crawl around the space on hands and knees and look upwards. It is then that the bison and horses come to life, flickering out of two-dimensionality, their muscles twitching, their eyes following wherever you go.

Cave-art specialist Derek Hodgson believes that rock art began in close observations of animals by hunters under stress. The shadows in caves are trigger cues, hints of danger. The cave environment and its tricks of light could become a simulation of the real hunt. As Hodgson writes, 'the visual brain of Palaeolithic hunters was not only specialized for detecting salient parts of animals, but became acutely sensitized to detecting animal outlines, which derived from the everyday experience of hunting'. Neural circuits were 'triggered' by the subdued lighting of caves and the risk of encounters with predators like cave bears. When these people saw animal-like shapes in the caves and rocks, they were painting over what a fearful imagination saw in the shadows.

There are other hints of this ancient attention. In a recent study on the amygdala, neuroscientist Florian Mormann found that when individuals are shown images of landmarks, people and other animals, the highest amount of cellular activity arises from looking at animals, whether a pet dog or a snake. In later studies, it became apparent that the animal response was in the right side of the amygdala. This apparent lopsidedness gives support to another theory that the right hemisphere of the brain responds more to unexpected but meaningful changes in the environment. As an animal ourselves, it is other animals – whether they're friends, enemies or something to eat – that get our brains and bodies going.

In making sense of cave art, anthropologists have turned to surviving hunter-gatherer societies that continue to paint inside caves, particularly the San peoples, who live in communities across a wide region of southern Africa. What began to fascinate anthropologists who studied the San was their meticulous imitations of the animals they hunt. The hunters, in some sense, become animals in order to make inferences about how their prey might behave. This spills over into ritual. The San use hyperventilation and rhythmic movement to create states of altered consciousness as part of a shamanistic culture. In the final stage of a trance, Lewis-Williams writes, 'people sometimes feel themselves to be turning into animals and undergoing other frightening or exalting transformations'. For anthropologist Kim Hill, identifying and observing animals to eat and those to escape might merge into 'a single process' that sees animals as having humanlike intentions that 'can influence and be influenced'.

Whatever sparked them, the cave paintings of our ancestors offer us mind-bending visions of humans turning into other animals and animals wandering about like humans: the Birdman

and the Sorcerer painted on the walls of the Lascaux cave, the Löwenmensch, a prehistoric ivory sculpture of a man with a lion's head discovered in a cave in Hohlenstein-Stadel in 1939. This is a world in which the minds and spirits of living things can be known and, in heightened conditions, exchanged. In this kind of shapeshifting, the hunter's essence can fly into his quarry and vice versa.

For Hodgson, the human visual brain is particularly adept at what is termed 'seeing-in', which concerns the ability to apportion agency to any suggestive-like form. The idea of seeing-in comes from some intuitions about human consciousness by British philosopher Richard Wollheim in response to Ludwig Wittgenstein's work. Wittgenstein wrote of 'seeing-as', where we can see something as the kind of thing it is. But with seeing-in, we can see something else in an image, like the famous drawing of a duck that might also be a rabbit. Wollheim's version is a demonstration of our ability to shapeshift what we look at. But humans are also mindshifters. Not only can we see other things in a single image, we can also see other minds in other things.

We shouldn't be surprised that the hunter-gatherer symbology ascribes agency to other animals. Hunters had to predict what

their prey would do in the future and outsmart any of those predators that might see them as a meal. Lewis-Williams believes that the religious view of these groups was a contract between the hunter and the hunted. 'The powers of the underworld allowed people to kill animals, provided . . . people responded in certain ritual ways, such as taking fragments of animals into the caves and inserting them into the "membrane".' This is borne out in the San. Like other shamanistic societies, they have reverential practices between human hunters and their prey, suffused with taboos derived from extensive natural knowledge. These practices suggest that honouring may be one method of softening the disquiet of killing. It should be said that this disquiet needn't arise because there is something fundamentally wrong with a human killing another animal, but simply because we are aware of doing the killing. And perhaps, too, because in some sense we 'know' what we are killing. We make sound guesses that the pain and lust for life we feel – our worlds of experience – have a counterpart in the animal we kill. As predators, this can create problems for us. One way to smooth those edges, then, is to view that prey with respect.

San cultures see animals as possessing special essences. In this, they are typical of most small cultures that live closely with wild animals. One idea is rendered phonetically as *!gi:ten/!gi:xa*, a supernatural potency found in the great animals, especially Africa's large antelope, the eland, which can pass into humans. Not only do other animals have skills and intentions, but these can also flow into humans, so that the skills become their own. In today's worldviews this would be branded as naïve anthropomorphism. But which view is closer to reality? The one that succeeds in hunting animals by assuming there's intelligence at work or the one that makes capital out of animals because they're mindless?

Today, we see into the experiences of the dogs or cats that

share our homes. In a Japanese study, Miho Nagasawa and colleagues found that oxytocin spikes in the bodies of both humans and their pet dogs when they stare into each other's eyes. Otherwise animals are almost exclusively understood as ostensibly mindless things for us to use without remorse. Even inside the conservation movement, individual animals or even populations can be replaced or killed without conscience in order to recreate some human ideal of biodiversity.

It is a curious turn of history that many of us obsess about machine intelligence while largely dismissing the evidence that other animals have a high level of conscious experience. Even those with quite different forms to ours nonetheless have intent. Something must account for why the members of large urban societies see the future of minds in pieces of machinery and code but no longer see them inside a mouse or a humpback whale. This is a kind of madness. Somehow, the notion of a unique human soul triumphed over a world of souls. And once the idea of a unique soul was doubted, the notion of the unique human mind cancelled out an abundance of minds.

One answer to this riddle may lie in a cave in Çatalhöyük in Turkey. The walls are adorned with dozens of men surrounding exaggerated versions of wild creatures, spears in hand. Around their waists are the pelts of Anatolian leopards. There's an almost complete absence of portraits of women, except for a handful who appear to be pregnant. This is a male world, in which the reverence once directed towards wild animals has apparently given way to taunting. The Shrine of the Hunters was found in the 1960s during the final year of the Mellaart excavations at Çatalhöyük. It is the site of the first relatively large settled township with domesticated animals, established over nine thousand years ago. But there are no images of domestic animals here.

This is a transitory moment in human history. The relationship of humans to other animals is altering. As the millennia pass and the domestication of plants and animals increases, archaeologists find that these ancient peoples depict fewer and fewer wild animals. The number of portraits of ibexes drops across the Neolithic period from thousands to a handful. The same pattern follows for all the wild creatures with which Palaeolithic hunters were once familiar. Instead there are images of dominance and slaughter, such as the killing of a whole herd of deer, as found in a mural on the walls of the Cova dels Cavalls. These shifts in image-making would seem to track shifts in human ideas about their relationship to other creatures. What might once have been viewed with awe were now tokens of human status. Other animals became rungs in the ladder of our ascent. And we can intuit that once animals are beasts of burden, cattle and wool, there isn't any need to think with the creatures alongside us. And possibly it is also an advantage to believe that other animals can't think. Perhaps when farming began to thrive ten thousand years ago, animals started to lose their minds. Or, rather, we began to lose sight of them.

There are a few anecdotes from the slaughterhouse that might support this possibility. Studies on empathy towards animals show lower rates among farmers and slaughterhouse workers. In the early part of this century, journalist Gail Eisnitz undertook hundreds of interviews. 'When I worked upstairs taking hogs' guts out, I could cop an attitude that I was working on a production line, helping to feed people,' one worker told her. 'But down in the stick pit, I wasn't feeding people. I was killing things. My attitude was, it's only an animal. Kill it. Sometimes I looked at people that way, too.'

In her postgraduate thesis, Anna Dorovskikh found that slaughterhouse workers experience a kind of stress-related response that she believes is based on participating in a traumatic situation. This builds on extensive work by criminologist Amy Fitzgerald. It's been documented for a long time that crime rates in America are higher in areas close to slaughterhouses, especially in relation to domestic violence and child abuse. The assumption has been that this is due to the demographics of the workers and employment rates in proximity. In looking at the workforce, there is a higher proportion of unemployed young single males, although fewer than in the farming community, as there is year-round pay. For years, the 'boom town' hypothesis of migrant workers led to under-theorising the consequences of killing animals. But Fitzgerald was the first to test these empirically.

Controlling for key variables, slaughterhouse employment was still associated with increased crime rates, unlike in comparative industries. Industrialised slaughter had the greatest effect. Other studies in different countries have since confirmed this. These discoveries can be hard to swallow but, in other ways, they can reassure. Whatever the ultimate causes, we get the sense that we are not an easy predator. In recent years, a range of studies have found that between 50 and 85 per cent of meat consumers would be unwilling to kill to eat meat.

But these kinds of dilemmas also creep into the work of

those dedicated to non-human animals. And how we think about the minds and experiences of other animals is key. In a study on veterinary students at Cornell University, researchers found that students were more convinced of the higher mental states of dogs and cats than farm animals. And this seemed to affect their willingness to perform certain procedures on them.

American psychology professor Hal Herzog has pointed out that we needn't get into the detail of how our treatment of animals is morally right or wrong. It's enough to realise that it's inconsistent. 'Moral codes are the product of human psychology,' he says. If the moral codes that arise from our psychology are inconsistent, they will affect us as much as other animals. And we can presume that the way we measure mental experience has profound consequences for us, too.

Humming smarts

Humans are only ever fleetingly conscious in the sophisticated sense of being self-reflective persons. More often, we rest in a state of perceptual consciousness that mightn't be hugely dissimilar to that experienced by other animals. Even more frequently, we pursue the many thousands of choices or decisions that require very little in the way of awareness at all. And a good third of our lives, we're asleep. None of this is to say that being aware of ourselves isn't important or immeasurably enriching. But this isn't the point. It is that we use this rich mental life to downplay the experiences and purposeful lives of other animals. In this, we don't have the truth on our side.

It is more than likely that conscious thinking of some

imaginable kind is one of the foundational roles of the central nervous system of animals. Randolf Menzel's study of insect brains concludes that no brain is 'structurally simple'. The honeybee, he points out, has 960,000 neurons at the latest count. Combine that into a hive mind of a low average of 50,000 individuals and you arrive at a collective brain power of forty-eight billion neurons. That's a lot of humming smarts. The next time you are walking through a forest on a sunny day, consider that the little ant at your feet has around 250,000 neurons in its tiny head. Then remember that an average colony has more than twenty thousand individuals inside it.

It's certainly fascinating that the domestic pig has much the same number of neurons as a dog. It's compelling that big land mammals like lions, bears and giraffes all have a similar and huge number of neurons inside their skulls. It's also relevant to our understanding of where we sit within the rest of nature that the numbers greatly increase once you look at primates like chimpanzees, orangutans and gorillas. The number we have more than doubles that of our closest relative, the gorilla. Although the elephant outstrips us all.

It is similarly interesting to look separately at the quantity of neurons in the forebrain, which seems to be highest among expressive, social animals. Once you look at these figures, dogs leap up the chart, alongside animals like parakeets, horses and ravens. The majestic elephant drops down behind the great apes. Instead, the whales swim ahead. It is possible that the long-finned pilot whale, a skilled group hunter, has more inside its monochrome flesh than we do. But what do these numbers really tell us?

The sea sponge has no neurons, no brain or nervous system. It's fair to say that this reduces its behavioural complexity. In responding to the needs of a sponge's life, our interactions will be profoundly different from those that would come into play in how we respond to each other. Yet sponges are extraordinary things. Their plantlike bodies can shelter thousands of other animals. Some live up to two hundred years on the twilit sea floor, indifferent to the philosophy of life but persisting like a quiet myth of their own. These animals have lived on the Earth for around a billion years. They have few predators. Outliving both pollution and disturbance, they are a token of the economy of resilience. Who are we to say they have no value other than as something to scrub the dirt off our backs?

Not so long ago, while I was waiting at a train station on a warm summer afternoon, I watched a single ant drag the lifeless body of a fly over a large area of stones in order to reach its nest. For nearly thirty minutes this tiny creature struggled with how to navigate this large burden – proportionally a bit like one of us pulling the carcass of a horse across a complex of massive boulders. I watched it come up against a sequence of obstacles. When faced with one of these, it would tuck the fly out of sight, come back up onto the surface and find a way round, then return to retrieve the fly and continue on its way. After an epic struggle, it made it back to the nest. Of what use is it to quibble over whether this is instinct or a more dynamic intelligence? There's something staggering about an entity smaller than the smallest nail on my child's finger problem-solving so impressively on a gravel pile by a backwater railway station.

We are surrounded by evidence that intelligence and awareness are a normal part of life. Instinct or otherwise, everything around us expresses an unquantifiable degree of complexity and skill. What prevents us from valuing it all more highly? Sue Savage-Rumbaugh has used simple keyboard skills in the Yerkes laboratory to enable chimpanzees to demonstrate that they possess the seeds of selfhood. Cambridge University scientist Nicky Clayton has successfully trained scrub jays to remember that certain foods become unpalatable after five days, a hint that episodic-like memory is possible even in animals without language.

When parrot expert Irene Pepperberg began her work on bird cognition, it was 'a crazy niche'. Now, she says, there's so much research on avian cognition, it's almost impossible to stay across it all. When she first began her work, the controls on studies on intelligence in other animals were too tight. The suspicion of the Clever Hans scenario – the miraculous horse who was actually responding to the unconscious cues of the researchers – made it difficult to get data. Even today, there's much less money put into research on animal intelligence because, as Pepperberg says, 'there's no obvious application'. But, after a lifetime of study, there's no question for Pepperberg that parrots conduct visual searches and have mental images.

One of her tests involves a tray with coloured pompoms covered by pots. After four swaps of the pots, she asks her parrots to find a particular colour. On average, humans get it right around 60 per cent of the time. Her parrot manages 50 per cent, and 100 per cent of the time if there's no swapping. But parrots aren't small, feathery humans. Like us, they're long-lived and order their world into social hierarchies. But they see the world in ultraviolet. They mate for life, singing each other special duets during mating season. Otherwise they live in huge flocks. But if the possession of mind is so valuable, how should we think about the fact that we have caused the endangerment of 50 per cent of parrot species, a quarter critically so? In crude terms, if we possess intrinsic value on the grounds of our intelligence, then surely parrots do too?

In 2009, Europe's top agricultural research institute published its findings that a 'high content of consciousness does occur' in other animals. It was followed in 2012 by the Cambridge Declaration on Consciousness, signed by dozens of leading scientists, confirming that 'non-human animals have the neuro-anatomical, neurochemical, and neurophysiological substrates of conscious states . . . including all mammals and birds, and many other creatures, including octopuses'. What is more, the authors state, they should be treated as such. Do they have a self-rationalising view like the one our social consciousness has given us? How far should this matter for us to recognise the worth of their lives?

Our concept of personhood gives us our beliefs in freedom of expression, free will, our right to education and liberty. These insights into the common wants and wellbeing of individual humans are both perceptive and sound. What they're not evidence for is the absence of meaningful needs or feelings in other animals. Nor are they proofs of our evolutionary superiority. As one of the innovators of CRISPR, biochemist David Liu, put it to me: 'Evolution is not a line, not a continuum, it is a tree. If you cut a slice across the treetops, the points across the plane make it clear we're not the end of evolution.'

This is pivotal when we face the challenges of large-scale losses of biological diversity as a consequence of our choices. In denying the feelings, intelligence or needs around us, we have left the possessors of these experiences to be expunged. We allow this because we believe we are the end. In a 2018 census on the world's biomass of life by Yinon Bar-On and others, it was found that the utilisation of marine mammals by humans has left us with a fivefold decrease in their presence since the start of human modernity. The total weight and quantity of wild land mammals was found to be seven times lower. Instead, around 70 per cent of the mammals on our planet are farmed animals we eat. The world's most common bird? The domesticated chicken.

Enter the algenists

The problem in the twenty-first century is that modern societies still act as if the old myths of human uniqueness tell us something true about the state of nature. In this techno-smart, science-obsessed era, we still base our legal systems on dreams of angels. This should concern anyone with a stake in reason. To navigate the social world of an animal like us, we deny the intelligence or value of those who threaten us or stand in our way. By some cruel irony, this means we can even deny the intelligence of our own bodies. Lewis Wolpert has proposed that the control of movement was the precursor to the brain. If the self-awareness we cherish began as images of the body to help in movement, it's an irony of nature that the most self-conscious of animals now doesn't want the body it evolved to know. In a weird extension of the idea of personhood, we can see our body as an animal that can be killed for the benefit of our minds.

There seems to be a part of us that resists the idea that there's something valuable about a sensing, physical state without a strong subjective experience. It feels powerfully as if subjectivity really is something abstract, a spirit of sorts. It's as if there is either an on or off switch in strong, conscious experience. Either a self or a great, senseless nothing. It goes against our intuitions to see our self-aware mental life as being crucially related to our animal bodies. Some of this we owe to the strange qualities of our own self-reflection. Many of us would find it hard to imagine tenderness or distress without some sort of self-consciousness. But some of this is down to the relative freedom our minds seem to enjoy in comparison with our bodies.

For a slave like Harriet Jacobs in nineteenth-century North Carolina, thoughts still wandered freely. Still today, in countries like Pakistan or Iran, the bodies of young gay men can be jailed but not their memories of each other's flesh. For New Zealander Nick Chisholm, who had an accident in his early twenties that left him enduring locked-in syndrome, it was torture to experience an unbroken mental life inside the motionless vessel of his body. 'It's an incredibly lonely existence at times. It's amazing how much time I have to think about things now since the accident. There's heaps of thoughts that I don't bother even expressing.' All of us feel that there is more to us than others can see. Those despised or misunderstood must feel this even more acutely.

In a more mundane way, we all undergo a feeling of disunion when a disease or virus strikes. Or when our bodies begin to age, and skin tags and bulges appear, muscles soften and droop, and the strength we once possessed diminishes, yet inside we feel as spry as a twenty-year-old. For all of us, in one way or another, it can feel a cruel fate to be an organic thing. Seeing the body as a prison, we long to break the locks and liberate the autonomous fly-by-night inside us. Despite this, we know that we experience pleasurable and sentient experiences without our sense of self exerting itself. There might be a lavish subjective experience with only mental images or even the invisible braids of scent.

By the late twentieth century, it had become one of our dominant fantasies to revise the human body through technology for the ultimate fulfilment of human potential. The Laissez-Faire Institute in Switzerland puts it that 'transhumanists and libertarians care about achieving technological progress, controlling nature, expanding the power of individuals over nature.' This worldview revolves around the intellectual capacities of people, the intention to upgrade bodily functioning and resist death. However, what these modern humanists seek to save is not our bodies but our minds. They dream of maximising what the mind can do and of lengthening the life of the body up to the moment when the mind can be rescued. These new forms of humanism are only secular theologies of salvation. Like all young religions, they began as experimental, radical and peripheral. Nonetheless, as with older creeds, our animal bodies are the enemy of deliverance.

In 1957, Aldous Huxley's eldest brother, evolutionary biologist Julian Huxley, coined the word for an emerging movement. 'The human species can, if it wishes, transcend itself . . .' he wrote. 'We need a name for this new belief. Perhaps transhumanism will serve: man remaining man, but transcending himself, by realising new possibilities of and for his human nature.' Today's believers have stayed loyal to this original view. Australian ethicist Julian Savulescu has written that biological enhancement is 'playing human', not playing God. 'I want to be the kind of human who lives longer and better, not shorter and badly.' He's not alone. James Martin was an infotech engineer, sometimes labelled as the boffin who predicted the internet. He was also a futurist who believed that we are 'capable of making the world enormously better'. He thought we should modify ourselves to be 'a better creature than we are today'. Philosopher Mary Midgley once called this way of thinking 'algeny', after the old medieval notion of alchemy. Such algenists long to turn the base metal of humanity into some 'undreamt-of gold'.

Another futurist, Ian Pearson, has pointed out that 'nobody wants to live forever at ninety-five years old', but if you could live forever in the body of a twenty-eight-year-old, many people would seek this out. But, in the various theories for how this might be achieved, there's no consideration of what it might mean to bypass childhood or reach maturity. Little thought is given to the presence of the menopause among social mammals like elephants and humans, whose elders have a key role in passing learning on to younger individuals. In foregrounding the reproductive stage of the human life cycle, we ignore the advantage of turning off reproduction to allow an animal to enter into a new role freed from the buzz of sexuality.

Woven into enhancement arguments is the notion of efficiency, as philosopher Nick Bostrom acknowledges. On Bostrom's website he writes of the 'limitations of the human mode of being' as 'so pervasive and familiar that we often fail to notice them'. Some of this idea of efficiency originates in mechanical and computational metaphors for human intelligence. Pamela McCorduck's *Machines Who Think* was the first history of artificial intelligence, written in the 1970s when the idea was really getting under way. For McCorduck, we have been ascribing mind and agency to our own inventions for millennia. She points to the idol worship across pre-Christian eras, and object worship too, whether totem poles or magical swords. The term for this kind of mind-swapping is anthropopathism. Through this, we saw spirit in any object made by a human. But there are two ways of achieving thinking objects – worship or trickery.

Artificial intelligence can be understood as both intellectual quest and search for salvation, as rigour and fairy tale thrown together with a dash of cynical entrepreneurialism. It was really in the eighteenth century that the founding principles of AI took root, in the efforts to build systems of thought. Today's obsession with AI is ultimately an offshoot of humanism and the debate about what persons are and how we can improve ourselves. In a final triumph, we can do away with the biological act of creation, the mess of women and wombs, and become our own makers.

McCorduck tells the story of Paracelsus and his homunculus. As the story goes, some sperm, sealed in glass and left in horse manure for a month or so, not only begins to self-fertilise but to take on the form of a human. This pellucid being must be fed artificially with human blood until, after the normal period of human gestation has passed, it comes to life. 'Not only will it live,' McCorduck writes, 'but it will have intelligence, and Paracelsus gives some detailed instructions on how it is to be educated. More to the point, he gloats, "We shall be like gods. We shall duplicate God's greatest miracle – the creation of man."'

South African entrepreneur Elon Musk has been one of several prominent technologists to express alarm at the possibility of strong artificial intelligence emerging in our century and posing an existential threat to human lives. In 2014 he likened the prospect of true intelligence in a machine to 'summoning a demon'. This isn't new. For decades there have been fears that something like personhood could emerge from zeros and ones, the dawn of minds in silico. Might artificial intelligence have its own intentions and might these be dangerous to our survival? In evaluating the risks of artificial intelligence, we place too much emphasis on the minds we might make and too little on the risks that arise from the minds doing the making. What would an animal like us do with any 'consciousness' that might emerge from our engineering of artificial intelligent systems? Given that we already rig the system to dismiss the intelligence of any being with inconvenient claims on our compassion, why would we imagine we can be trusted? The more likely danger may flow from us to AI, rather than the other way round.

Still, stoking such fears can lead us to overemphasise the advantages of machining our own intelligence. In summer 2019, Musk reported that his company Neuralink had created artificial threads that might enable a technological prosthetic to the human brain. As is often the case with controversial research, the general public is presented first with the promise of freedom from a frightening or debilitating disease. The 'neural lace' is to be pioneered to alleviate certain neurological conditions. But the eventual goal is to provide us with the option of becoming a cyborg, half-human, half-machine, in the event that artificial intelligence outsmarts us and we need a boost in brain power.

From a metamorphic gift for seeing into minds, we have been left with a philosophy that urges us to engineer our own minds. At the same time, we've come to see our future in the computers we once invented as a parlour trick. And now we see these shadowy analogues as a possible threat to our own status. Like an ouroboros, the logic bites its own tail. To save ourselves from a superior intelligence that we fear will view and treat us as we've viewed all other animals, we're told we must become machines ourselves. But this dash to engineer ourselves risks diminishing the experiences that make our lives a pleasure. So, too, might we unintentionally remove the critical parts of our social behaviour that remain our best hope of anything that approaches virtue.

And so, if we no longer see into the lives of other animals, it's not because they don't have minds or we can't. It's because we don't want to. Yet now we're told that everything should make way for humankind's greatest invention: artificial intelligence. This is a far more dangerous delusion than anything dreamed up in a church. In this cult of freedom, nothing much is said of the consequences for the eight million or so species that live alongside us. Little is said for the passing of all the intelligence found in flesh and bone, feather and fang.

The last frontier

For those who believe that our subjective consciousness is what saves us from the insignificant lives of the rest of the species on our planet, technologies that might simulate this are seen as a tool of salvation. If persons are ultimately only sensory impressions in the brain, they can be restored or even uploaded. The persons who have true value can become the last frontier of technological mastery. Manufacturing a person has become the final goal in the long investigation into how thought arises from biological matter. But it has merged with the parallel inquiry into how we can be saved from death.

Those caught up in this goal look for events that take place in the brain because they seem to offer a more promising way of living forever by living without our bodies.

In one sense, we might think about persons as bodies that have come to know themselves. The first self may have been nothing more than a basic image of the body to assist in the navigation of the world. Other mental images would help in distinguishing a friend from an enemy, food from a predator. Some of this recognition might take place through scent and taste. But visual cues are often critical. As Antonio Damasio reminds us, the human mind has its foundations in 'bodily feelings, primordial and modified', and, as with all organisms, its business is 'the successful regulation of life'. Yet it's possible to believe in today's major cultures that the body doesn't matter a great deal. The somewhat embarrassing body, with its whims and its wobbly bits, is edited out of the picture. As philosopher Derek Parfit put it, 'the body below the neck is not an essential part of us'.

It's easy to see why. When we concentrate, our own thoughts are so absorbing that we can momentarily forget that we have a body. Then something brings us back. A pang of hunger. A noise. And this expansive, floating projection of thought cascades through the body again and into a whole awareness of ourselves and the room. This is both bewitching and more than a little odd. Our intuition tells us that we are not really the creature of muscle and bone that stares out from the mirror. We are the conscious thing in our heads. In this way, we don't have to believe in dualism to be under the illusion of it. We are trapped in a sensation of personal experience.

This sensation has been tackled by philosophers with some far-reaching consequences. In the latter half of the seventeenth century, English philosopher John Locke devised a thought experiment to expose something a little spooky about human identity. In his experiment we have a prince whose soul, upon his death, enters the body of a cobbler, such that the prince opens his eyes one morning to peer out from the flesh of another man's reality. The challenge this idea sets us is that most of us intuitively feel that the cobbler has become the prince. This seems to provide proof that we are a mental rather than a physical thing. Our mind is a thing of spectral beauty and integrity that could as happily reside in a piece of manufactured silicon as in a mythical winged griffin. Like the contents of a suitcase, the thing we are can be emptied out and packed in again elsewhere. Following this, if a person is a mental substance, be it a soul or *res cogitans* or a piece of computer code, then under certain circumstances anything can be a person.

Locke's thought experiment is a riff on our natural propensity to believe that living things have a hidden essence. There's a related piece of work from the 1980s by Yale psychologist Frank Keil. He studied how young children understand natural categories. When asked if a porcupine that has woken up one day as a cactus remains a prickly animal or should be thought of now as a prickly plant, children routinely maintain that the cactus is still a porcupine. But the same would not be true of something non-living like a car or a table. If the car woke up as a cactus, it would no longer be a car. What Keil's experiments show is that it is intuitive for children to think in terms of living essences.

Locke knew, as we do today, that humans are multicellular organisms, marvellously self-organising forms whose underlying biology is conserved throughout life until the death that leads on to decay and the irreversible change of form. But Locke grew up in a religious culture in which serious concerns about the rapture were a common part of theological discussion. The

body parts of saints had been scattered around Europe, distributed to separate churches, traded and stolen. How would such
body parts come together again in the event of the rapture at
the end of days? In answer to this, Locke began to think about
the ability of the mind to shift between the copious, almost
inexhaustible memories of the past and experiences of the
present. He became convinced that this time-travelling ability
created the unique identity of a person.

Locke's ideas have been profoundly influential. It's still his
musings that make sense of why a country like Saudi Arabia
gives citizenship to a robot while continuing to suppress the
rights of its own flesh-and-blood women. If a person is a set
of rare mental skills, then we can get humans that aren't a
person and humanoids that are. Saudi Arabian women have
been denied legal status as persons on the basis that they lack
sufficient capacity to conduct their own affairs. Similar prejudices
justified the anti-suffrage movement in early twentieth-century
Britain. As Lord Curzon put it at the time, women were without
'the balance of mind' to make sensible decisions. Much the
same was argued against black suffrage in America. At the 1821
New York State Constitutional Convention, Samuel Young
asserted that 'the minds of blacks are not competent to vote'.
Fighting for equality and equity in terms of human persons of
any kind has been a necessary move against prejudice.

But, in a weird twist, some of us, mostly in the countries
that first fought to elevate the concept of personhood, now
delight in the idea that persons aren't real. It has become a
strangely attractive possibility among secular individuals that
the very thing we believe makes our lives important is nothing
but neurochemical flimflam. Here is the marriage of strong
reductive beliefs with an ancient breed of dualism.

One of the favourite proofs comes from American scientist

Benjamin Libet, who died in 2007. Libet's experiments seemed to show that the decision to act comes milliseconds before conscious awareness. His experiments have been taken as proof that one of the central tenets of personhood – what we call free will – is a mirage. Never mind that free will and the means of conscious choice might not be the same thing. Studies like this have perturbed and fascinated. They seem to show that our perception of personal control is just a trick of the brain. And this notion of trickery is reassuring to those who want proofs that our minds are some kind of simulation in the brain. This way, persons can be engineered for our purposes. If we can figure out what personal thoughts are and how to decode them, we can write them again on our own terms.

Admittedly, some of those who speak of the 'self-illusion' are actually arguing against the idea of the self as a separate thinker of our thoughts. This is fair. The sensation that we are a 'thing' carried along by our body isn't a biological fact about us. But the laudable effort to puncture such a view is undermined by popular declarations that selfhood and free will are only brain-based illusions. Such individuals end up arguing one absurdity to destroy another.

The real problem with the response to Libet's proofs is the notion that an unconscious, embodied mechanism of cognition is a threat to autonomy. And the idea that animal behaviour is illusory is a consequence of the mistaken belief that there's some neat divide between instinct and rational choice. For those that hope a person is only conscious thoughts in abstract, rather than the thoughts of an animal that feels or touches or gives birth, there's nothing at stake in swapping what we are into an eternal form. Not only that but we can write ourselves into synthetic forms that are safe from all we dislike about being a creature. If it's all a big simulation, we needn't worry about running the simulation elsewhere. There's nobody whose authenticity we might violate. If it's all an illusion, why not upgrade to an illusion that lasts?

Of course, we don't yet know if we can take the brain apart and find the bits our conscious thoughts are made of. Our consciousness may be the wild enigma of a path up that has no path down. And few who are wonderstruck by studies of the brain stop to consider what a strange idea it is that an illusion can be extracted. It's like the scene in *Peter Pan* when Peter's shadow is cut from his body by the closing of the window through which he has climbed, only to be kept in a drawer like a length of cloth to be stitched back on later. If the self is a shadow cast by the light of experience, what will be stitched into the materials of a machine?

All that our experiments have done is show us that we are a creature, woven through with the brilliance of instinct. It's only when we forget that our conscious experience is a feature of our bodies that we stumble when we see it at work. When we swallow too much literature on how humans are not animals, it can then surprise us that consciousness or selfhood or rationality are a physical part of our lives. That our subjective consciousness is a physical phenomenon that can be interrupted by everything from diet and disease to depression only reaffirms that we are an animal. Pain, too, can be thwarted by interfering with chemical signals and the brain, but we don't tell the soldier who has had a leg destroyed by shell fire that their experience is an illusion.

It takes the briefest examination of ourselves to see that consciousness of every process or behaviour that our bodies must perform would cripple us. Consciousness, whatever else it is, is necessarily parsimonious. That awareness of action trails unconscious instinct doesn't make it an illusion any more than it makes the body an illusion. Nor does it render our awareness dysfunctional or incompetent. Rather, our sense of self is a purposeful extension of the needs of our bodies.

The modern angelologists

The great minds of medieval Europe once spent a fair bit of their time struggling to make sense of angels. How could an angel, an ethereal being without eyes or ears, recognise Jesus's mother Mary among the other women of Galilee? For Thomas Aquinas, the solution was in the ability of angel consciousness to pick out the unique signature of a person's thoughts. Aquinas didn't consider the small matter of whether angels might be figments of human imagination. Angels had something irresistible to say about animal life. If disembodied entities like angels were above humans in the ranks of beings, then intelligence didn't need a body. The evolution into a superior state lay in the abandonment of flesh.

These days, we have our own latter-day angelologists. Roboticist Hans Moravec sees a future of 'exes' – short for ex-humans – existing in a post-biological world. 'Physical activity,' Moravec tells us, 'will gradually transform itself into a web of increasingly pure thought, where every smallest interaction represents a meaningful computation.' Such flesh deniers are everywhere now. For a futurist like Giulio Prisco, the grand frontier of space 'will not be colonized by squishy, frail and short-lived flesh-and-blood humans . . . It will be up to our post-biological mind children.' Our enlightenment, these men claim, lies in the freedom from our animal bodies.

Similar views now motivate whole research programmes and commercial enterprises dedicated to the notion that we can somehow exist without being animal. The embattled Human Brain Project, a multi-billion-euro Swiss research initiative, alongside Hewlett-Packard, was conceived to build a whole simulation of the human brain in a supercomputer. 'The ultimate goal should be to model the unique capabilities that distinguish humans from other animals.' Around the same time, Russian media mogul Dmitry Itskov founded the 2045 Initiative with the aim of life extension for humans in the form of computer avatars. The aim is 'to create technologies enabling the transfer of an individual's personality to a more advanced non-biological carrier'.

Not to be outdone, Google's Larry Page launched the Calico life-extension project, while its head of engineering, Ray Kurzweil, announced a few years ago that a 'shot full of nanobots will someday allow the most subtle details of our knowledge, skills, and personalities to be copied into a file and stored in a computer'. Even the physicist Stephen Hawking stated that it is 'theoretically possible to copy the brain onto a computer and so provide a form of life after death'. As a man whose body had forsaken him, we can hardly fault him for this.

Posthumanists are the intellectual heirs to ancient dualist ideas that it isn't our bodies that make us who we are. An offshoot of humanism, these thinkers see the human condition as nothing more than a technical issue we've not yet solved. Today's posthumanists are convinced that being human is neither here nor there, so long as personal consciousness is present. Individuals like Itskov and Prisco say to hell with the body. We can reboot the personality in an immortal form or project it out into space like human thought gone wireless. For them, the animal body is a problem best done away with. Where once humanists were concerned with making sense of how matter becomes thought, now posthumanists are concerned with how thought can escape matter. It's an exotic twist on an old tale.

It should be noted that there are other posthuman projects of a very different character. Pioneered by philosophers like Rosi Braidotti, these belong more properly to visions of a world after humanism. These posthumanists aren't seeking to get past being animals but to get past narrow humanist ideas about what humans are. In the words of Elizabeth Grosz, they are rethinking a world 'beyond the limited perspective of human self-interest'.

Yet, despite such re-examinations, it remains a potent dream that our minds can save us from being animal. Driven by capitalism and libertarian fantasies of personal autonomy and frontiership, our experience is to be engineered for survival. Given enough time, we can be free of flesh. The problem of being animal will be solved forever. Some vocal proponents revel in the talk of computer simulations of human intelligence inheriting life from the unfortunate flesh-and-blood creatures that evolution has kicked out. But the attraction of machines and algorithms as evolution's next stage is a stark reminder that it's fear and not rationality that moves us.

Voltaire once counselled wryly, 'The human brain is a complex organ with the wonderful power of enabling man to find reasons for continuing to believe whatever it is that he wants to believe.' In seeking to extract our personal perspective like an oyster from its shuck, these posthumanists ask us to believe that something that has irreplaceable value can be replaced. This is a reckless idea. They fatally undermine the very thing they hope to save.

In any case, experiments and trials in the real world must

first be undertaken. For this, animals would be used and, later, humans. In the modern age, this means grant applications and ethics committees and a set of legal parameters. Historically, we have justified experiments on animals because they don't qualify as persons to whom we might owe anything. In parallel, we have justified the need for the research and innovation because the flourishing of human persons is thought to be a moral act. But, for this, we must assume that human persons are real things to whom we have moral duties. How then can we argue that they are only physical illusions that we should seek to replace?

It's likewise a problem that such discussions are typically expressed in vague progressive terms. In a book on Alan Turing's legacy, Ray Kurzweil wrote of 'the freeing of the human mind from its severe physical limitations of scope and duration as the necessary next step in evolution'. In his view, evolution moves towards 'greater complexity, greater elegance, greater knowledge'. It's a sentiment that is echoed often these days.

Advocates of technologies that see the next stage of human evolution as the abandonment of our organic bodies for an artificial and upgraded intelligence bring together an assumption that matter explains mental experience and that mental experience is mostly a set of problem-solving rules. But they also borrow from the old philosophical dream that the conscious thinking self is the quiddity of true value. In this way, they say, algorithms can be turned into the soul.

What Kurzweil has in mind is evolution on a diet. Using only one aspect of evolutionary processes as a grounding principle is like using flow to justify a river. Darwin never made such a mistake. At its heart, evolution is about dynamic interactions, not progress. Life reaches its endless forms by blind genetic mutations and recombinations, some of which become more common because they help the individual in their lifetime. Differential survival and reproduction result in the physical changes to species. But the changes that emerge aren't in one direction, nor are they about improvement.

Evolution can both create more complex organisms, which looks a lot like a direction, without creating better organisms, which would give direction a purpose. Evolutionary changes can be staggeringly slow, as in the rare goblin shark, a pale mash-up of a creature that has been much the same for over one hundred million years. It lives in the deepest parts of continental slopes and seamounts, swimming slowly, eating when food swims too near. In its twilit world, there's little to trouble its quiet life. Other changes can be rapid and substantial but largely invisible in their effects. Dart frogs in the Amazon were separated over a period of thirty years by a motorway. They don't look different but they're genetically distinct.

It's never easy to tell from genetic mutation when speciation occurs or if a marker results in a new behaviour. It's even less clear what a new species, more complex or otherwise, has to say about the relative importance of living things. 'In three thousand years,' Tanzanian anthropologist Charles Musiba told me, 'if some future humans travelled back and saw the Mbuti people [pygmy hunter-gatherers of the Congo region], would they see them as less important, less evolved?' Whatever else it is, evolution isn't a process of enhancement. In dressing up salvation in evolutionary language, posthumanists use science to justify prejudice.

What's more, our biology biases our thinking in some critical ways. Locke wished to see a person as made from their relationship to memories of the past and imaginings of the future. When people think about personal identity, treasured and even painful memories make up much of the content. Yet when people consider what they want to save of themselves, they tend to single out the stage of a healthy adult. What people favour is an adult of reproductive strength. It takes little imagination to recognise why we might prefer this phase of life. But what does this mean for the aged, the young, or those with different kinds of perception, such as people on the autism spectrum or with Down's Syndrome? Presumably they aren't worth saving in the same way.

A similar set of dilemmas beset early theological debates. If people were to be resurrected at the end of days, what stage of a person's life would return? Would the flesh reanimate or only the spirit? Would we stay forever as the age that we died? Endlessly, it's the animal body and its life cycle that causes the muddle and our stubborn longing for the supposed best days of our lives.

Rainbows of flesh

It's a strange experience to sleep and wake into a dreamed room that is identical in every way to the one in which you went to sleep. The same curtains, the same bed. Every detail the same, only it's dreamed, a brain-figment, a large, physical room miniaturised to electrochemical pulses inside the brain that rests inside the skull that rests on the pillow. This room is unknowable, unvisitable. Yet to the sleeper, it is fully realised and experienced. It has texture, space and sound. It is in all ways as fulfilling a world as the one experienced before the sleeper lay down.

And what if the sleeper walks through the room, unaware she has a sleeping double? She'd laugh at the suggestion she is nothing but a dreamed phantom, an electrical wraith inside a lump of matter. What a ridiculous idea! And yet, at any moment, she will be extinguished by some noise or hum of light that brings the true sleeper out of her dream. The dreamed being will be erased instantly, leaving no trace, nothing.

I once had a sort of dream in which I lay awake in my bed while an exact copy of myself dressed, pulled on the light in the bathroom and stepped down into the darkness of the stairway. Once fully awake, I then rose from the bed and into the pathway of this dream like a car pulled onto the treads of a prior journey. It was as if for a moment I had cheated time and glimpsed a

future dull and ordinary but a few seconds ahead of the reality
in which I stood. Who was the more real? The future self? Or
was I suspended not in the present, as I had assumed, but in
some past state? A figure on a mechanism, carved by some
invisible hand, tracking a preordained route. The memory of this
half-dream still unnerves me now. By some glitch or another, I
felt as if I'd caught sight, however briefly, of the backstage
workings behind the grand façade of the present moment.

In a letter written to Lord Kames in 1775, Scottish philosopher
Thomas Reid asked a question that has since become known as
the duplicates problem. Reid wanted to know if his brain died
but somehow, several hundred years later, could be revived and
fabricated in a new form, 'whether, I say, that being will be me;
or, if two or three such beings should be formed out of my
brain, whether they will all be me, and consequently one and
the same intelligent being'. In other words, is there continuity
or does the original cease to exist and a new one replace it?
This matter of duplication is a serious business for those who
seek to be the architects of our escape from animal life.

More recently, philosophers Andy Clark and David Chalmers
came up with the extended mind thesis, which is based on the
'active role of the environment in driving cognitive processes'.
When Chalmers talks of an extended mind, he includes the
human body and some of its environment. These kinds of
philosophical discussions can get pretty trippy. Imagine in this
extended mind that we replace a biological part with an external
piece of kit that does some of the job of thinking. It's still
participating in the thought processes, yet it's no longer human.
As Chalmers writes, 'A subject's cognitive processes and mental
states can be partly constituted by entities that are external to
the subject, in virtue of the subject's interacting with these
entities via perception.' This theory allows for the whole body

and environment to be part of our mental life, but also for the possibility that biology might be inessential.

Thought experiments allow philosophers to try to think through what constitutes mental life and how it might be categorised. But they are theory only. While some of the body's thinking might be translatable into synthetic parts, it isn't proof that this is true of all parts. After all, some parts of the body are more essential to phenomena than others. We recognise that we can remove a limb and life may continue without any decisive change in who we believe ourselves to be. But it is equally possible to imagine a point at which a critical level of bodily disruption either transforms subjective experience entirely to make a different kind of person or – at some point – snuffs out the person altogether.

In dicephalic twins, two persons exist in one body with two heads. Abigail and Brittany Hensel, joined at the torso, have spoken about how one of them can be hot while the other is cool to touch, and one of them can experience a stomach ache on the other side from where their head is located. 'Isn't that funny?' one of the girls once remarked. Such twins have distinct personalities and can sometimes be separated successfully. This seems to support our intuition that there is a separable person inside us that matters, for we can alter their bodies seemingly without impunity. What is important is that the consciousness of the twins survives.

But it is more disorienting when we consider the rare blend of craniopagus twins. Born in 2006 in Canada, Tatiana and Krista Hogan are joined at the head. The girls also have separate identities, but their experiences are shared in astonishing ways. The girls have two separate brains, like dicephalic twins, but they are also coupled by what their neurosurgeon calls a 'thalamic bridge', an arc of tissue stitching them into one mind. One girl loves ketchup and the other refuses to eat it, yet both taste it.

They know if the other is being touched, even if their eyes are closed. To an extent, they know one another's thoughts without having to speak. 'Talking in their heads', they call it. Krista and Tatiana have a pooled individuality. For them, the private owner-ship of conscious experience is different. There's nothing odd or freakish about their experience. They are two girls in a flesh rainbow. But does this mean we can look forward to a future in which an engineered version of their thalamic bridge might free consciousness, like a maple tap releasing the sap from its tree?

All this is fine when we're simply throwing around some ideas. It's a little different once the debates enter the operating theatre. In the last decade or so, an Italian–Chinese team of neuroscientists has been attempting to innovate a form of head transplant. For Sergio Canavero, the self is 'simply an illusion that can be manipulated at will'. Canavero assumes 'all aspects of that which we consider uniquely ourselves reside in the brain. The body is simply a support system.' He and his colleague Xiaoping Ren have experimented on dogs and monkeys, and the bodies of dead humans. Most recently, they claimed triumph after transplanting a new head onto the existing body of a living rat. The work has met with a mix of anger and mistrust, as well as accusations of releasing results publicly before peer review. In some ways this is typical of medical frontiers. Not so long ago heart transplants provoked anxiety and even repul-sion, where now they are largely accepted. And if the person we are is just a mental sleight of hand and the body just an elaborate walking stick, what's there to worry about? Nevertheless, it remains an under-discussed reality that organ transplants carry with them a considerable psychological burden.

Concetta De Pasquale's studies of kidney transplant patients found that transplantation is profoundly gruelling, even while it offers life. The new organ disrupts the whole system of the individual, from the firing of neurons to hormonal changes. Transplantation can result in a 'psychosomatic crisis that requires the patient to mobilize all bio-psycho-social resources during the process of adaptation to the new foreign organ'. This, De Pasquale says, can lead on to changes in 'self-representation and identity, with possible psychopathologic repercussions'. In the months after undergoing surgery, a patient may experience what is called 'confusional psychosis' along with feelings of anxiety, hallucinations, disorientation and sudden mood changes. What might follow if it is an entirely new body, rather than only an unseen organ?

The dreams of whales

What we seem to have forgotten in the present age is that our most basic assumptions about the world around us remain speculations. This isn't because the methods of science are faulty. Science has given us ample evidence. But certain things elude us. Perhaps, too, we have certain things wrong. This shouldn't surprise us. History is full of wrong ideas. But these gaps in our understanding can be quickly occupied by human desire, which fuses almost invisibly with what seem like reasonable ideas. For nearly a century we have thought of the brain as an electrical, logical computer plonked on top of a meaty machine. It's to the computer we look to make sense of what we are and what the body does. But these kinds of ideas may one day be viewed in the same light as those of the leading physicians of Europe, who used to cut apart frogs to establish the workings of the soul. The same intellectual vertigo might well grip readers of neuroessentialists, those that believe the brain is in some sense the equivalent of concepts like the self or the person, in another fifty years.

Most scientists don't believe in dualism, as neuroscientist Alan Jasanoff notes, 'at least not consciously'. But, he says, the classical intuition that mind and body are somehow categorically separate has lingered. One problem is in the popular assertion that 'we are our brains'. The other problem is the metaphor of the mind as software. For the past few decades, we've based everything on algorithms as rules for problem-solving. Computers have become exceptionally powerful by relying on this form of intelligence. Yet none of them has shown even a flicker of the awareness of a tortoise or any other living creature. One possible reason for this is that consciousness may depend more on neurophysiology – a multi-directional weft of cells, nervous systems, hormones. It's much more to do with the 'wet and slimy things' of our bodies, says Jasanoff. He encourages us to remember that our brains achieve their information transmission because of electrochemical signalling through the interactions of neurons. But around half of our brains' cells are glial cells, which aren't electrically active. These non-neuronal cells form part of the central and peripheral nervous system, and are believed to play a role in communication and neuroplasticity.

In the last decade or so, it has come to light that there is constant communication between our brains and our guts, through a tangle of chemicals and cells. The gut is so big that it has its own intrinsic enteric nervous system, which then communicates with the central nervous system. Fundamental to the development of the enteric nervous system are the neural crest progenitors, migratory cells that colonise the gut as a baby grows in the womb, wrapping around the boundary tissues of the digestive tract. Millions upon millions of tiny organisms in our guts – what we call microbiota – have a crucial role in the whole wellbeing of our bodies, in the release of inflammatory mediators, vitamins and nutrients that support essential processes like the immune system and metabolism, as well as the regulation of brain function. Allan Goldstein, a surgeon at Mass General Hospital, has dubbed the gut our 'second brain'.

Studies of the microbial influence of our guts on emotion or behaviour have been growing over the past decade or so, suggesting a potential role in depression and schizophrenia. Studies that have taken gut bacteria from depressed individuals and placed them into the guts of mice have reported that the mice display changes in behaviour that are similar to depression. Other studies have shown a high occurrence of gastrointestinal symptoms among those with autism spectrum disorders. In searching for consistent patterns in different personalities, researchers have found that traits like conscientiousness and neuroticism are affected by higher quantities of proteobacteria, like helicobacteria or salmonella, in our guts. People with so-called 'leaky gut', where higher quantities of bacteria and toxins pass through the intestinal wall, report changes in their personalities, especially anxiety, as well as systemic immune responses such as inflammation.

Research to show what the body does has been practised both on rodents and other primates. In 2018, a new technique was showcased in which dead mice can be made as transparent as plastic to gain an exceptional insight into the cellular inter-actions of the body. The technology is called vDISCO, offering glimpses into the complex ways in which brain injuries impact on the nerves and immune system elsewhere in the body. The dead mouse is soaked in solvents and stripped of fats and pigments. Green, fluorescent nanobodies are flooded through the circulatory system of the tiny body, which then glow under a microscope. The images are a vista of the glittering synergism of an animal body. They show us what we face in unpicking the mind from the flesh.

*

In June 2018, while I was studying in the US, I took my family to Woods Hole Oceanographic Institution on Cape Cod. There I met up with Colombian neuroscientist Rodolfo Llinás. Over the past few decades he has researched the squid giant synapse, the largest chemical junction in nature. It is fundamental to the jet-propulsion system that thrusts the creature through the deep seas of the world. Housed in the alien-like fleshy mantle of the animal, the giant synapse has given us a model for how all chemical synapses function.

We must acknowledge, he said to me, that we have only a loose grasp on what we mean either by intelligence or by consciousness. In Llinás's view, consciousness can be split into two separate definitions: the state of being aware of and responsive to one's surroundings, and the phenomenological sense of being aware, in which there seems to be a person experiencing all this. Intelligence is the ability to apply knowledge or skills and it can be either knowing or instinctual. Intelligence is in all living things: 'It is a property of cells.'

Some kind of awareness allows living things to interact intelligently to survive. When multicellular systems integrate, there has to be some kind of internal organisation. But the ability starts to narrow down. While plants receive information that brings about changes to the plant, the nervous system of animals is a specialisation. Once you have a distributed nervous system, the need to order movement causes a brain to appear. For Llinás, some kind of consciousness may be needed to coordinate complex movement in an animal of multiple moving parts. This may be related to Purkinje cells, branched cells in the cerebellum that regulate movement. In the 1980s Llinás and colleagues discovered P-type calcium channels within the Purkinje cells of mammals. P-type channels, it turns out, are crucial to the nervous system and the heart, and critical to the release of hormones at the ends of excitatory or inhibitory synapses.

When we seek to define what matters about us, we fall back on the feeling of subjective consciousness that we possess, but this can blind us to a more complex reality. We still don't know how consciousness occurs, although it involves an enormous number of intracellular properties. The brains of the animals that have emerged on our planet share universal features. Given the high degree of similarity between the bodies and brains of multicellular life on Earth, the principle of parsimony, that things most often behave in the most economical way, suggests that at least some animals have subjective experiences. In some ways, Llinás argues, subjectivity is 'what the nervous system is really all about, even from the most primitive levels of evolution'. Consciousness is just a convenient word that stands for a global function that emerges from but extends beyond our immediate anatomy. The person we build our lives around is a consequence of the body, a detonation of senses and interpretations, the meaningful content of the central nervous system.

Whatever is going on is both immensely complicated and deeply embedded in the whole body of the organism and its environment. In mammals, the central nervous system is connected to every part of the body, a weft of nerves so long it could take you halfway across the Earth. It includes the gut, the ganglia, the length of the spine. And our anatomy isn't just tailored to our needs, but we are tailored to our own planet. Humans have the foramen magnum, a hole in the centre of the base of the skull for the spinal cord. The elegant reason for this is that a bigger brain would cause the head to hang. In human species, as the brain size of our lineage grew, the hole moved forward so we could walk. We marvel at our capacity to travel to the Moon, but we don't appreciate the fact that we can't walk once we're there. Away from our planetary home, subject to different gravitational fields, movement and thought would be different. Our brains and bodies are marvellously adapted to the properties of the planet on which we were born. We're earthbound creatures and consciousness is 'substrate dependent'. In other words, as far as Llinás's lifetime of research suggests, it is unlikely that we can separate consciousness from the body in any uncomplicated way.

The 'ability to stay alive at all levels of the system' in organic forms is what will limit consciousness in machines. Intelligence, yes. 'The liver is intelligent,' Llinás told me, 'so is the heart. They work beautifully but they have no self-awareness.' In this way, it is easy to understand why we might create intelligence in a machine, and certainly why we can simulate awareness without having it. But there's something off-key in the modern-day anxieties about machine intelligence. Llinás doesn't believe consciousness will be created in inorganic forms.

While we were chatting on that sunny June day, Llinás's colleague entered the lab to begin work on the machines behind us. The good-natured interruption shifted our conversation onto dreams. Llinás joked about how he tells his grandchildren that he 'dies every night' when he goes to sleep. In truth, there is something like consciousness present during REM sleep but not non-REM. Most land mammals experience REM sleep, when dreams are likely to arise, as anyone who has had a sleeping cat on their lap can attest. Research on some whale species in captivity suggests that whales may also undergo these intense phases of sleep. Hanging fifteen metres below the ocean surface like puppets on invisible strings, what images stir in their resting minds? Why are we not more curious about it? If we don't yet see the potential of consciousness in other life forms surrounding us, what makes us think we'll know it in a machine?

Llinás was in his eighties when we met, his curiosity undimmed. What is similar to consciousness without being consciousness? What kind of physical process creates a person's awareness? Does it emerge like sweat is secreted? Or is the mechanism more like the generation of tension in muscle fibres? Somewhere there is crucial information missing. His hope is that someone, someday, will understand consciousness and treat the knowledge with the respect it's due. 'If I knew that someone really understood the issue of consciousness,' he said to me, 'I could die without knowing myself.'

A STRANGER TO CREATION

Whereas all beings have their place in nature, man remains a meta-physically straying creature, lost in Life, a stranger to the Creation.

Emil Cioran

Panic, pathogens and predators

A few years ago, exhausted from nursing my young son, I was out walking in the forest where I was living. It was a bitterly cold day. Needle-ice slit the damp soil, leaving small shining masses, as if the winter had laid her eggs. I felt happy enough. But in less than a moment a shade cut across. My heart began to beat like crazy and my palms began to sweat. It was as if the uncontrollable elements of the outside world had passed into me and obliterated my sense of self.

What I'm describing is a panic attack. This kind of sudden terror is an animal shadow sweeping over the human form. It's as if something wild has only been sunken under the outermost layer of rationality. When, for whatever reason, there is nothing to keep it back, it escapes, made volatile by its confinement. The senses reel and then sharpen. Ordinary experience is suddenly outclassed by the hidden senses of the body.

Someone who is panicking irrationally in this way is full of a jumbled narrative to try to pull themselves back to a state of self-possession. Reassurances, *it's okay, it's okay*, are swiftly negated by a convincing feeling that the world's about to end. And while the brain's language centres work at any narrative that might assist, other regions of the body bang out norepinephrine and serotonin to prepare for an imaginary threat. Legs become twitchy and get ready to run like hell. Yet, before long, the panic begins to retreat again. The discharge of fear vanishes back into the privacy of being. Anyone who has endured a moment of groundless panic like this chides themselves for overreacting, then swings into doubt again. Finally, the emotions begin to dull. In their place comes a more painful mood, one of humiliation and wonder. What was this strange, uncontrollable encounter?

Some people will recognise this experience instantly. Others may not. But, even if a person hasn't gone through a panic attack, we are all aware, each in our own way, of the mysterious, disconcerting regions of selfhood. Whatever it is that triggers a feeling of panic, whether a stressful time of life or nothing at all, the form it usually takes is a hypersensitivity to the sensations of the body and the impression of being trapped inside something that will die. On that wintry day, I found my heart racing with unfathomable anxiety. But I realised something that should be blindingly obvious, once the fifteen or so minutes had passed. For the first time, I saw why my body was frightened. It was frightened for itself.

Being an animal is frightening. In common with all organisms, our bodies can be harmed, from unwanted sexual advances to physical injury and stress. Many of us don't like what we discover about our animal bodies and the perils they face, from the discharge of faeces to the cancerous proliferation of cells.

But physical threats like pathogens and predators have profound effects on all living beings. And animals have a vast range of methods to beat the organisms or viruses seeking to get a hold.

When we're busy at our spring cleaning, it's worth sparing a thought for the wood cockroach, which defecates in its own home because its faeces have antifungal properties. Some beetles roll around to coat their bodies in antimicrobial mud. Corsican blue tits line their nests with lavender to ward off mosquitoes. Termites smoke out dangers by fumigating their nests with naphthalene.

So, too, the need to outrun predators has left our planet's life with an awe-inspiring collection of behaviours. One of the best-known predator defences comes from the bombardier beetle, capable of internally mixing together its own dynamite-like concoction of chemicals to blast away a threat. The beautiful word 'crypsis' is used to describe the spectacular ability some animals have to resemble their environment. It's a life-saving skill. In a classic study in the early twentieth century, naturalist Alessandro di Cesnola selectively placed some brown and green mantids among a mix of vegetation. Over a period of eighteen days, birds systematically gobbled up all of the insects that didn't match their background plants.

But of course competition and aggression take place between

individuals of the same species, too. All manner of tricks to outwit can be found. Primatologist Dorothy Cheney told the story of two captive male chimpanzees, Luit and Nikkie. The pair had been skirmishing for dominance over a protracted period. On one occasion, Luit attempted to hide his involuntary grin of fear and submission by pressing his lips together with his fingers, out of sight of Nikkie. Only once he'd got his expression under command did he turn back to face his opponent. And strategies to overcome threats will vary, not only between species but among individuals within a species. In a study by biologist Dietrich von Holst, tree shrews showed different reactions to a lost contest. One response was to become submissive and get on with things, but others seemed to give up.

As we might expect, our bodies also have powerful strategies for avoiding dangers. The sensations of thirst and hunger help us prevent dehydration or starvation. We have senses that allow us to select and digest the right kinds of foods. Our immune systems offer us an internal retaliation against the threat of pathogens and other agents of disease, and we also have psychological traits like the experience of disgust, which turn us away from possible sources of contamination like rotten meat and foul odours.

All animals show the evolutionary advantages of an efficient threat-processing circuitry. Even protozoa exploit the ability to detect light in order to sense danger. But in an animal that has an evolved capacity to be apprised of and, to a certain extent, to regulate some of its experiences, we arrive at some truly tortuous and versatile behaviours. We are conscious of what every animal wants to avoid – disease, injury, an inability to persist. Being conscious doesn't mean that we're thinking about it all the time, but that we're, knowingly or otherwise, aware of the threats involved in being a living organism. This awareness of the hostile actions of the world around us exerts a hefty and often invisible influence across our lives.

American neuroscientist Joseph LeDoux began his work on rodent responses to threats, but he rose to prominence after the 1977 publication of his book *A Divided Mind*, which reported discoveries about the human brain that he and Michael Gazzaniga made almost by accident. When working with patients that had undergone hemispherectomies – the removal of half of the brain – LeDoux and Gazzaniga discovered that we identify a possible danger in a split-second of unconscious perception but that a separate part of the brain builds a conscious narrative to make sense of the fear. They uncovered this when flashing a dangerous or disturbing image at a patient too fast for them to register what they were seeing, measuring the fearful response in their body by detecting a raised pulse and sweating palms. Next they asked the individual why they were afraid. Despite being unable to recall what they'd seen, the patients gave elaborate, unrelated explanations for their fear.

LeDoux was in the middle of writing 'a deep history of humans' when I met up with him on a warm afternoon in Williamsburg, Brooklyn. In recent years he has argued that altered threat-processing is at the heart of anxiety disorders. He believes that anxiety should be understood as two systems, rather than one. There are the behavioural responses and accompanying physiological changes in the brain and body, and there are the conscious feeling states reflected in self-reports of fear and anxiety. These are the sorts of entanglements of perception, instinct and autonoetic processes – mental and imaginary time-travelling – that are specific to humans. There's a tension in our bodies, not because our kind of cognition isn't a natural event but because we want to make sense somehow of sensations that are often hidden from conscious analysis. And also, perhaps, because we are privy to things we'd rather not see.

'Fear is the cognitive assessment that you are in harm's way,' as LeDoux tells it, 'a view that allows fear to arise from activity in any survival circuits or by existential concerns.' We react to predators and protect ourselves from injury and pathogens. But we also recognise threats, those to our status for example, which are subtly different. There's still an adaptive cause, but the consequence is at a remove. Originally, status was about priority access to resources in situations of competition. Reductions in status might lead to a life-or-death scenario, sending alarm bells through the whole body of an animal. Particularly for male primates, disease, death and ageing can precipitate a disastrous loss of rank and therefore reproductive opportunity. Today, shifts in our social status still result in measurable differences to our serotonin and androgen levels.

At present, people with anxiety disorders are treated by targeting the amygdala, the ancient, emotional segment of the brain, but with only partial success. The reason for the focus on the amygdala is because of cases where people with damage to this region of the brain fail to exhibit bodily reactions to threats. For LeDoux, this is an oversimplification of what is going on. There are other aspects of the brain that we are still only beginning to understand. It is known that the ventromedial prefrontal cortex regulates the amygdala, altering or suppressing a threat response as the situation changes. The failure of this mechanism to kick in can result in an outbreak of uncontrolled anxiety. After the initial split-brain experiments, Michael Gazzaniga went on to develop his theories that the left brain acts as an interpreter, making sense of and justifying responses after they've happened. Gazzaniga and colleagues lit upon a part of the left hemisphere that is implicated in narrating a unified experience. He nicknamed it 'The Interpreter'.

There is a tendency in studies of humans to seek a primary reason for this or that behaviour. But all that we have discovered so far should counsel us away from simplified causes. Much of human cognition seems to happen in a kind of commotion, in which a range of internal systems and processes collide, as well as external influences. This is the difficulty we face in making sense of ourselves. What goes on in our conscious minds is a maze of reactions between evolved impulses, open-ended behaviours, external influences and memories. What good can come of trying to straighten out a species whose narrative mind is a concoction of astonishing apprehension and equally astonishing confusion? When we try to reduce our species to a simple explanation, we find ourselves tumbling in and out of lucidity, as if some distant controller is fiddling around with the channels.

Gazzaniga is clear that reductive explanations of the brain fail to capture the vast matrix of processing elements that make up what we understand as mental behaviour. This makes straightforward evolutionary explanations difficult, as only some of these processes will have clear adaptive properties. Others may simply be swept up into the tempest because many of our characteristics have been co-opted for multiple reasons. This also makes us a rather limited creature when it comes to engineering straightforward purposes to any of our traits, but it is essential to what gives us such behavioural complexity. The dynamic interplay of our bodies and our ideas is a dramatic feature of human experience.

But, in appreciating that humans have intrinsic responses to threats that we share with most other life forms, we can see that our interpretation of risk, when combined with memory and the real-time social environment in which we find ourselves, makes for an eccentric relationship with the original threat. Our variety of consciousness, endlessly trying to make sense of itself, must do something with feelings of danger.

Unfortunately, we no longer acknowledge the original sources of many of our feelings or impulses. Our social intelligence has had the effect of screening them out. Even more vigorously, most of us live in societies that convince us that to be human is to be a rational individual in complete control of our choices. Yet we see glimpses of an alternative reality in our studies of how we respond to threats. In psychopathology, researchers have long associated many of our most common mental disturbances with a defensive analogue of some kind. British clinical psychologist Paul Gilbert has suggested that paranoia maps to the threat of hostile others; hypochondria and obsessional disorders to the damage of one's body or the contact with something harmful. Social anxiety maps to the consequences of lowered status; separation anxiety to the loss of sources of protection or support.

Even though this is perfectly intelligible to most of us, we still prefer to deny that we are an animal. We believe our cultural and intellectual lives have freed us from natural causes. But we forget at our peril that the way we see the world and each other remains a function of the social psychology of a primate.

Better together

The shift to sociality among many kinds of animals is regarded as one of the major transitions in evolution. Animals, from a species of butterfly to baboons, can come together for better access to potential mates and shared ways of getting food. But some of the explanation for group living lies in the upper hand individual animals gain over threats. The collective noun for zebras is a dazzle, the strobe-like effect of a herd running together to bewilder a predator. The flocking of birds may reduce the so-called 'domain of danger' in the absence of any other hiding place.

But evolution doesn't unfold as a neat sequence of difficulties that are overcome by the survivors. It draws on a cluster of entities and events from which nothing we might strictly call success emerges. Each evolutionary gain is vulnerable. While for social animals it's mostly better to be together, groups also generate novel problems or dilemmas. One problem is competition. Another is that close living gives parasites, viruses and bacteria a fantastic opportunity to spread.

When a new coronavirus virus emerged from the Wuhan province in China in late 2019, epidemiologist Mark Woolhouse told journalists that 'no one is surprised the next outbreak is in China or that part of the world'. The reason is that Wuhan is a city of high population density, with around eleven million citizens. It is disruption and proximity that create the conditions for zoonotic diseases. Closeness is the pathway to transmission. Of course, we now know that this virus, SARS-CoV-2, led on to the devastating pandemic of the disease COVID-19, which shut down societies and economies around the world and killed hundreds of thousands of people. Pandemics have always affected human populations, just as they do other species of animals, whether they originate in influenza viruses or bacteria that give us diseases like malaria or bubonic plague. What is of note now is that most emerging infectious diseases of the past century or so have arisen through intensive human utilisation of other life forms and their habitats. Sometimes these plagues come from our appetite for meat. Sometimes they come from our destruction of the places occupied by other animals. Once the spillover of a pathogen from another animal has occurred, the spread is then aided by our high degree of connectivity, which can circulate a virus around the world in a matter of days.

Disease-causing agents can wreak havoc and misery on animal populations. Unsurprisingly, then, there are all manner of ways that animals in groups complement the individual defences of their bodies with extra methods of protection. Social immunity is the term for a new phenomenon, defined originally in 2007 by evolutionary biologist Sylvia Cremer as 'the collective action or altruistic behaviours of infected individuals that benefit the colony'. At first the effect was studied almost exclusively in insects. Among eusocial insects, different collective mechanisms thwart some of the canny ways in which parasites or pathogens succeed in infecting the members of a colony. Some species remove corpses from nests. Others isolate infected individuals. Several ant species have what might be called instinctive graveyards, special refuse chambers for nestmates. *Temnothorax lichtensteini* ants remove dead related ants from the nest. But they bury the dead bodies of unrelated ants. And these sorts of group defence mechanisms can also be seen in animals that have only a fleeting phase of social behaviour, such as *Melanoplus sanguinipes*, a migratory species of grasshopper whose gregarious individuals have an adaptation to avoid eating prey infected by certain kinds of parasitic fungi.

A species of subterranean termite prone to infection by fungal spores turn their bodies into an alarm system on infection, vibrating to warn others. Termites from the same colony then box the individual in so that they can't infect other members. And always there are arms races. *Aethina tumida*, a small beetle found in sub-Saharan Africa, parasitises honeybees by feeding on young bees and other food sources in the hive. The adult bees herd any beetles they discover into little prisons made of tree sap. But rather than starve to death, the beetles have adapted to use their antennae to tickle the adult bees guarding them, which mimics one of the ways bees feed one another. Sometimes they can trick bees into offering them a meal of honey.

Recently, the research of Sarah Worsley and others revealed the arms race between leafcutter ants and a parasitic fungus called *Escovopsis*. This fungus has found its niche by growing on the food source of a leafcutter colony. All being well, worker ants forage for fresh leaves on which the fungus *Leucoagaricus gongylophorus* grows because it breaks down the plant into an edible meal for the ants. Over time, the *Escovopsis* fungus has specialised in exploiting these little gardens. And, over further generations, the ants' bodies have evolved forms of fungicide, while the fungus has countered with internal chemistries that block these defences. And so it goes on.

In 2015 the idea of social immunity was refined by Joël Meunier, who sought to redefine it as any mechanism 'at least partly due to the anti-parasite defence it provides to other group members'. Among primates, the classic example of social immunity is 'allogrooming', the time spent picking parasites out of one another's hair. A hangover from this can be seen in the pleasure men and women take – to say nothing of the money they spend – in the hairdresser's. But, for humans, the possibilities for supplementing our intrinsic immune systems through coordinated actions are astonishing. Cooperation can give us everything from penicillin to better sewers. Within weeks of a

new virus like SARS-CoV-2 emerging, we had international agencies to offer advice, internal protocols, laboratories to analyse the new threat and publish their findings on how to mitigate or eliminate it, and a global infrastructure of hospitals, doctors and medical supplies. Hospitals, in the most fundamental sense, are external immune systems that people have built.

The methods by which animals work together to hold off dangers elide into the related idea of social buffering. This is the phenomenon now found to exist in a wide variety of animals in which the presence of other members of the same species can act to reduce the stress experienced by an individual. Being together can make some social animals, like us, feel better too. In a study by primatologist Frans de Waal, grooming between macaques was found to lower heart rate. Highly social mammals recover faster from stressful or dangerous situations by being together. Conversely, members of species that behave in this way can experience extreme levels of stress when isolated.

In related work on rats by Jason Yee and others, methods of social bonding in young animals reduced the effects of stress on the body and correlated with longer lifespans and lower levels of certain diseases. These benefits have since been found in animals as diverse as bats, lions, horses, meerkats and deer.

When squirrel monkeys see a snake in their environment, their small bodies release the hormone cortisol. The prominent endocrinologist Hans Selye was the first to try to make sense of stress-hormone responses in the 1950s. The hormones are released to prepare our bodies for danger, but they come at a considerable cost. This built on the work by Walter Bradford Cannon. When animals confront a potentially life-threatening danger in the wild, they undergo something we've come to call the fight-or-flight response. This state of hyperarousal originates in part in the adrenal medulla in the amygdala of the brains of vertebrates, leading to a waterfall of hormones that speed up the heart, slow the digestive system, constrict the blood vessels in the body and free up energy for the muscles. It's the fight-or-flight response that we experience when we have a panic attack.

The hormones of our sympathetic nervous system that act on threats underpin much of what we refer to as 'emotional experience'. They shift the body towards those things in the environment that might feed or stimulate. They make our hearts thump and our skin break out in sweat. But the impacts of these physical responses can be detrimental. To mitigate the effects, we and other animals have a range of defences against stressors, including the release of gluco-corticoids, studied by Selye, which protect against the body's reactions to stress, helping to prevent the loss of homeostasis.

In making sense of social buffering, what we now know is that if the squirrel monkey happens to be accompanied by a member of its group, the amount of cortisol released in response to a threat is lower. The same is true for us. And these effects are partly learned or at least altered by upbringing. When squirrel monkeys are raised away from their families, they don't show the same social advantages as their peers who were raised by their mothers and their own social group. For squirrel monkeys and for humans to gain the benefits that come from being together, individuals need to learn the supportive response through those first crucial relationships in early life. If we don't experience protective or nurturing relationships in infancy, there's less of an opportunity to gain the benefits of a manageable response to stressful events.

For humans, social support makes a considerable difference to the release of cortisol in our bodies while we undergo chal-lenging experiences. It also confers cardiovascular benefit. Such is the importance of good relationships to our overall physical

health that psychologist Mario Mikulincer and his colleagues have found that awareness of dangers, especially existential threats, increases the efforts people make to begin connections, from friendships to romance. Israeli social psychologist Gilad Hirschberger has also found that fear boosts the time spent on romantic relationships.

Think for just a moment of the sorts of things we do. Every day, we touch one another: a handshake, a friendly pat on the back. After a stressful period, many of us will go swimming, stroking our limbs through the water. Or we might book a massage and lie quietly while someone rubs their hands rhythmically across our skin. If we don't have the resources for these luxuries, we might sit curled up with one another at home. Our children roll around together, laughing. How often do we place our hands on their heads or smooth their skin, or bend down to inhale the particular smell we find in their hair, their necks? It's easy to take this for granted, but it's a precious part of the animal experience of those creatures like us that thrive in societies. For hundreds of thousands of years, our ancestors have generally been better off with one another than without. We might have forgotten it inside our cultural convictions, but our bodies haven't.

The solace of shared dreams

I was a student in the UK when 9/11 happened, but I went on to spend a lot of time in New York later in my twenties. I remember my first visit to Ground Zero vividly. I arrived at Fulton Street station and walked in the direction of the ruined site, which was still boarded up. It was surrounded by closed-circuit cameras and hardnosed security officers with orange vests and construction hats. The main viewing areas were crowded with tourists licking ice creams and staring blankly into the empty expanse before them. Several signs asked the question *What's Going On Here?*, explaining the phases of redevelopment occurring beyond the eyes of the onlooker. Close by the signs were white posters that read *Please donate to the Memorial. Visit www.buildthememorial.org*, and an emboldened message, **It's Time**. Despite the warning against unofficial posters or messages, there were names and dates and photographs. There were hearts, stars and other symbols with names of the lost written in every kind of ink inside them.

I stayed for a while, and then the delicate jingling of bells in the distance caught my attention. I took the first left, entered a quieter, shadier block where the bells were clearly audible, and followed. Ahead I saw a tree, glinting in the sunlight. Chimes hung from its branches, catching in the wind, pirouetting on air, rags of paper fluttering from the smaller branches and leaves. I picked one to read. Later that night, I wrote the words into my diary. 'To flee from memory . . . Had we the Wings

. . . Many would fly.' I had recognised it at the time, but I couldn't remember where the lines came from. It only took a moment's search on the internet the following day to find it. It is a poem by Emily Dickinson. Her last image is of birds watching the flight of 'men escaping/From the mind of man.' It gives us the impression of people flying away from what haunts them in their own minds, like balloons released by a distracted child.

The tragedy of 9/11 – and it doesn't have to be this singularity of ideologies, motives, aggression and despair but any other human tragedy from any other time or culture – is a peculiarly human phenomenon. No gorillas or jaguars are mounting suicidal and murderous attacks on one another for the sake of what flashes in their imagination. In this we are unique. But when bewildering events of human violence take place, we still cling to explanations that keep us squarely in control. We look to economics or ideologies or to the lessons of history. We don't so often turn to the story of our bodies. But those in power have long known what the rest of us are less willing to admit. Humans are animals whose responses to threats are predictable enough to be exploited.

Like other social animals, we find safety in numbers and we find comfort in our relationships to other members of our group. Yet it's more complicated for an animal with our kind of intelligence. For humans the common threats to our animal lives are beamed through the prism of an unusually flexible mind and interpreted in some truly creative and extravagant ways. As a social animal, our worldviews bring us together and provide us with the solace of shared dreams. We groom one another with visions of salvation. We massage one another with promises of our invincibility. And we defend these visions to death because they save us from death. And so, while all animals seek to overcome threats, humans have psychological tendencies to defend against threats to the ideas we use to overcome threats. To put it another way, we fight and, sometimes, kill ourselves and one another for ideas that are supposed to protect us.

Some intriguing studies from the past few decades have found that when we see danger, real or imagined, most of us seek to reinforce whatever worldview or culture we believe we belong to. Studies have since shown that threats can also lead to a stronger desire to meet the norms of the group. Harvard psychologist Carlos Navarrete and colleagues designed a series of investigations in which participants contemplated scenarios that contained an adaptive challenge. In six studies conducted in the US and Costa Rica, Navarrete's work found that participants who contemplated a range of threats that might be eased in a social group all increased their support of a pro-nationalist author over a social critic when compared to those primed to think about more neutral subjects.

Some researchers investigating our behaviour have uncovered evidence of a relationship between direct threats and the prevailing system of ideas. In a controversial study across thirty nations, the overall presence of parasites across a community correlated with measures of political ideology. Countries with greater burdens of infectious diseases were more likely to have authoritarian regimes and to be more religious and collectivist. Researchers also found that those with higher sensitivity to the potential of pathogens showed stronger regard for traditional norms. This research should be treated prudently. It doesn't mean that these ideological choices are directly caused by pathogens or our sensitivity to them. Instead, it seems that pathogens and how we react to them may be a factor in making us more or less inclined towards our group. Even if the effect is small, this might be significant in a future of megacities with populations of tens of millions living in close proximity.

Other research has shown that fearful events or threats can cause us to reaffirm our commitment to the groups with which we identify. Patriotic ideas like nationalism seem to increase our sense of security. Studies on this phenomenon have been repeated across cultures. Psychologist Hongfei Du from Guangzhou University found that a mechanism to achieve some degree of relief from threats by committing to traditional values exists across all cultures, but its exact complexion differs depending on the norms of the given culture. Impacts on wellbeing matched the ideals of collectivist cultures like China, while those of the Austrian participants in the study more closely aligned with Western individualistic cultures.

What follows from this is that sometimes a challenge to one's group or worldview can lead to feelings of jeopardy, with some extremely unpleasant results. Indirectly for humans a threat to our culture or worldview can feel almost like an attack on our immune system. Psychologist Jeroen Vaes believes that fears or perceived threats may also prompt people to see in-group members as more distinctly human. In effect, this can leave outsiders to be perceived as if they have fewer of the human capacities that warrant a moral response. In studies on Israeli participants, Gilad Hirschberger found that existential anxiety caused people to assert the superiority of their belief systems over others'. This can be achieved by attempting to convert followers and by derogating those who persist in challenging their worldview. 'In particularly ghastly cases,' psychologist Matt Motyl writes, 'people will support or participate in the killing of those who threaten their system of meaning.' In another study, Jeff Schimel found that having their cultural worldview threatened made people more fearful about death. The effects remained even when controlling for the emotional state of the participant. In the extreme, when our culture seems to be in danger, some of us may feel it as a danger to our own lives. Paradoxically, a few individuals will even sacrifice themselves for the sake of a group whose ultimate purpose was to keep them alive. So it is that, for us, the positive benefits of camaraderie can turn domineering or even violent.

Since the events of 9/11, there has been an extensive research effort to track its effects on politics. In 2001 a national survey of adults and children in the US found that more than 90 per cent experienced one or more symptoms of stress related to the attack, with the majority saying they sought relief by turning to religion, to friends, and to the groups with which they identified. Subsequent surveys, once some time had passed, found that these symptoms mostly faded away, except among those in proximity to the events. But a number of studies from the mid-2000s through to a recent survey in 2018 have found that the majority of those directly affected by the attacks underwent a shift to more conservative politics. The theory goes that people are motivated to act defensively when faced with threats to the legitimacy or stability of their belief system. In 9/11, the threat was to the idea of America itself.

'Conservative' is a term that is host to a wide range of different political and historical implications. For now, we can take it to mean a strong preference for traditional values most often found in the dominant worldview of a country. Conservatism, in this less politically charged interpretation, is attractive because it presents us with a world that seems straightforward and stable. A disputed but thought-provoking paper by George Bonanno and John Jost found that the 'conservative shift' among World Trade Center survivors in the months and years after the events was associated with a desire for revenge and increased militarism and, at the same time, with elevated levels of depression and stress. This isn't because conservatism makes people depressed. In fact, studies Jost has conducted more recently have found elevated levels of self-assessed happiness among those who describe themselves as conservative. Rather, it's the other way round. The shift to a worldview that feels more solid and conclusive follows the intense jolt of fear.

A plague upon thee

When we are in the business of interpreting everything through the lens of political history and culture, we deny ourselves some valuable sources of insight. When we push the reality that we are animals out of view, we obscure what might be motivating our actions. Of course, people get extremely sensitive about any hint of determinism in their political or religious affiliations. Certainly, a good dose of caution should be retained when interpreting these kinds of findings. But we can nonetheless be confident that we use our responses to threats in the service of our ideas.

Psychology professor Jamie Goldenberg has studied the basic disgust reactions towards animals, as well as the human body, its products and processes. These are part of our individual immune responses to disease. But these disgust reactions seem to increase when we're fearful or under stress. In a study from 2001, one group of participants was asked to write about their own death and another about a control topic. They were then asked to complete a disgust sensitivity questionnaire developed by Jonathan Haidt and others. When it came to bodily products ('You see a bowel movement left unflushed in a public toilet') and animals ('seeing a cockroach'), feelings of disgust were higher among those prompted to consider death. These innate mechanisms can be manoeuvred for more sinister purposes by a social being like us.

The use of disgust-promoting metaphors against those we wish to avoid or even attack is universal. When conflict and hardship led to a phase of increased migration into Europe from countries in the Middle East and Africa in 2015 the *Sun* newspaper described Britain as 'plagued by swarms of migrants and asylum-seekers'. Even the then British prime minister David Cameron called the crisis a 'swarm of people coming across the Mediterranean'. Those that might have inspired sympathy were re-categorised as viruses and pests. This kind of talk is frightening.

But contemplate a future world in which there are resource shortages and several billion more of us than there are at present. Consider also what might happen if climate change adds to large-scale migrations. In 2017 alone nearly nineteen million people were displaced by natural disasters. A wave of mass migration, somewhere around 10 per cent of the world's population, took place in the decades leading up to 1914. Another phase happened after the upheavals of the Second World War. What might unfold if a sequence of threats collide? We might expect there to be temporary or even extreme forms of worldview defence as populations and ethnicities mix. We can already see hints of this in the rise of nationalism and populism in the countries of the world's old democracies.

In 2018 the inventor of the World Wide Web, Tim Berners-Lee, acknowledged that 'Humanity connected by technology on the web is functioning in a dystopian way. We have online abuse, prejudice, bias, polarisation, fake news;

there are lots of ways in which it is broken.' In research by the World Economic Forum, it was noted that the goals of so-called 'cyber-racists', through group websites and forums, are to disseminate racist propaganda and, in turn, to make their own group stronger. In the WEF review, group prejudice online was found to follow remarkably similar patterns regardless of the ideals of the group. Methods to attract people included exploiting humour to manipulate how information is received and the selective use of news to fit their purposes.

And it helps us to know that certain kinds of leaders are particularly adept at exploiting our fears in order to increase commitment to their authority. For those high in the structures of power, threats are a tool to rally their populace. In 1984, Ronald Reagan released a short film called *The Bear*. In the film, a grizzly bear prowls around a forest, a symbol of the threat of the Soviet Union. It formed the basis of his election campaign and the justification for increased military spending. In 2004, while America was still living under the shadow of 9/11, the political campaign of George W. Bush released an advert that showed wolves gathering in a forest. In the final days of a tight battle between Bush and presidential hopeful John Kerry, the wolves were supposed to symbolise the threats to national security that Americans faced if Kerry was elected. The election, Bush told supporters, would determine the 'war on terror'. In the advert, a narrator tells the audience that liberals and Democrats cut spending on intelligence and defence in the years after the attack on the World Trade Center. Kerry's running mate, John Edwards, responded immediately, telling the attendees at a rally in Florida that the opposition were now 'using a pack of wolves running around a forest trying to scare you'. But it worked.

More recently, in an analysis of the Arabic Twitter accounts of the terror group ISIS, political scientist Sara Hamad Alqurainy and colleagues found that certain formulas were consistently used to incite violence against Westerners or other Muslims with dissenting opinions. The formula starts with distancing tactics that include the use of propaganda that describes 'the targeted group/person as subhuman, inhuman, bad' and as incapable of the warmth or decency of fellow humans. It quickly leads to justifications for the path to violence.

A study of dehumanising rhetoric used by recent presidential candidates in America found that other candidates' dehumanising comments ranged from zero to eight over the total election cycle, while Donald Trump had 464. While there's no question that there is movement of illegal migrants across the US border from Mexico, Trump has repeatedly described such unknown individuals as 'animals' and an 'infestation'. Such rhetoric acts to alarm the listener, reducing empathy for the wider situation and encouraging citizens or voters to intensify their commitment to their own group identity. This is significant, whether you're for or against a leader's policies. If you have an authority acting on people's fears in this way, it's very difficult for voters to be confident about whether they're being manipulated.

The exploitation of our fear systems and social tendencies needn't always involve derogating outsiders. Sometimes commitment to authority can be accomplished by reinforcing the superior skills of a group, culture or nation. In recent years, Chinese authorities under Xi Jinping have funnelled money into a wide-ranging programme of activities to promote the achievements and ideals of the communist government and to undermine detractors. Journalists Louisa Lim and Julia Bergin undertook an extensive survey of the evidence, finding that methods to support Chinese exceptionalism involved everything from censoring unflattering depictions of history to takeovers of media outlets. The hope, as expressed by Xi, is to 'tell China's story well'. This kind of narrative of state superiority is typical rather than unusual. So-called 'soft propaganda' has become a customary feature of state events throughout the world. But there's a fine line between pride and coercion.

And these tendencies, never edifying, can sometimes be dangerous. If we see the actions of others as in some way threatening to our worldview, we can seek to suppress people. By seeing their choices – everything from sexual preferences to belief in a different God or a different political candidate – as threatening, our antipathy towards them is legitimised. What's more, if we have dehumanised them, however subtly, we are much more likely to deny such individuals the full minds and feelings of fellow humans. And if we can induce a feeling of disgust towards their way of life, we can even find ourselves justifying cruel ways to subdue them. When fears are high, we should watch what we say.

The human myth

Modern civilisations have sought to escape the narrowness of our tribal tendencies by turning humanity into a universal source of empathy and compassion. This was certainly the aim of many social reformers and legislators after the horrors of the Second World War. This progressive ideal has relied on concepts of uniqueness that presuppose a natural system in which our relative status is the highest, simply because we are a member of *Homo sapiens*. But using the ultimate hierarchical idea to mitigate the dangers of hierarchical thinking is like using a mallet when a hammer's too clumsy. What it's useful to remember is that the idea of being human has mostly been about belonging to a group and not to a species. This presses on us whether we are willing to face it or not.

These days, the preference for an ideological belief system that reassures us that we are unique and, in effect, superior living beings is so pervasive that it can be exploited in research. The Kteily measure looks at graphical descriptions of the ascent of man, depicting the physiological and cultural evolution of humans from apes to full evolvedness. The image manipulates the underlying assumption that we are evolving progressively, and that modern humans are at the pinnacle of creation. But how sensible is it to use a false premise? Does it bring out the best in us?

The most immediate weakness is that it turns our reality into a threat. What we are contradicts what we tell ourselves we are. Once we have a hierarchical worldview that denies the facts of being an animal, then anything that suggests otherwise can undermine that worldview. Whatever reminds us we're animals might be displeasing or even offensive. In certain circumstances, this could include our own bodies and those of other animals. Even the title of this book – *How to Be Animal* – might induce a defensive response in some readers, for instance.

There are some signs of this defence against our animal reality in our reaction to the idea of death. Heidegger once said: 'Only man dies. The animal perishes. It has death neither ahead of itself nor behind it.' Something like this has been argued for centuries. While some, like Jacques Derrida, have come back with the possibility that humans perish too, as noted wryly in his book *The Animal That Therefore I Am*, they have been few. Death for many people is understood as a fundamental assault on who we are. Yet there is consolation in the belief that humans can be noble in their ultimate end. Meanwhile, death is some unseen snip in the lives of other creatures.

In many ways that view has been demolished over the past few decades. In 2017 a young boy from Arizona captured some footage of collared peccaries, a species of New World pigs, nudging, sniffing and visiting the dead body of one of their group. The behaviour continued for ten days straight, until coyotes finally got the better of the vigil and made off with the corpse. Primatologist Sarah Blaffer Hrdy has reported that she was mobbed by langurs while trying to reach the corpse of one of their babies. She is one of many who work alongside other apes and monkeys and bring us accounts of the common practice of carrying a dead baby. Others offer stories of juvenile primates refusing to leave the side of a dead parent.

In 2018, the internet was briefly lit by the images of a Pacific Northwest orca who continued to carry her dead calf for seventeen days. The researchers working with this community of whales pointed to the fact that they are an endangered group whose food source, the Chinook salmon, is also endangered. The pod had failed to raise a calf successfully for over three years. 'The whales have a very strong drive to look after their offspring,' said Canadian orca specialist John Ford, 'and this evidently extends to neonates that die at birth.'

Of course, it's not a revelation that death might matter to living things. Or that death matters to us humans. After all, many of us find it frightening that we will one day die. Both our breath and heartbeat will slow up and stop. The muscles of our body will lose their grip as if unpegged from the world. The jaw will fall open, skin will sink towards the ground, some of the body's excreta will pass without effort. The body will be utterly still and cold, until the moment of rigor mortis, when our eyelids may curl back like a singed corner of paper. Whatever we once saw of the world will be gone.

Only the other day, my young son, after being tucked into bed, tiptoed quietly back into my room and stood at the entrance, his face hanging in the darkness of the corridor like a moon. 'I've just been thinking how strange it is,' he said, 'that we're alive.' 'Yes,' I replied, 'I know, love. But it's time for bed now.' Of course, this was hardly a satisfying response, but it was the end of a long day. He lingered. 'Do you think any of me will stay behind when I die?'

Primatologist Thomas Suddendorf has described the mental time-travel that he believes gives humans their unique form of conscious experience. He writes movingly of where this temporal human imagination can lead. 'As a child,' he says, 'I struggled to come to terms with the most unwelcome of all realizations: the fact that, one day, I would die. I lay in bed, staring at the ceiling, and attempted to imagine "not being." I thought I might be able to do this, as the state seemed little different from dreamless sleep. However, I couldn't get my head around the idea of never existing again – never to wake up; to be gone forever. Even now the thought makes me feel queasy.'

Death is always there in the background of human life. There are deaths from diseases as run-of-the-mill as pneumonia and as nasty as Ebola. There are homicides, suicides, fatal acts of domestic violence, car accidents, famine, civil wars like those in Yemen and Syria, and deaths from less visible causes like pollution. But, most of the time, these deaths aren't made salient to us unless they touch us directly. Most of the time, we forget. Death is largely kept from view in modern life. Of course, certain events in human history can change that, like a war or a pandemic. A lethal virus, invisible to the human eye, can make visible all that the human mind doesn't wish to see. Death, grief, the passing of time, the temporariness of it all. So powerful are such ideas to human minds that we are affected when confronted by them, in ways both small and large.

Over several decades researchers across a range of fields of study have built up a huge body of work on our responses to death. What they have shown is that death through the prism of a human mind becomes curiously distorted. While, for some, mortality is a waking horror, for most other people it is a buried anxiety. There is considerable evidence for this fact beyond the common sense of one's own experience. Even in its most extreme forms, anxiety affects a significant number of people; by some calculations it is the most common mental disorder in the world. But milder forms of worry, fleeting irrational fears of disease, a sudden preoccupation with a sensation in the body, an outbreak of feeling about death, are almost universal. What is more, much of the anxiety that people have unfurls in the body without conscious engagement with the original source. The anxiety remains even if we have powerful reasons to be reassured of an immortal afterlife.

In a 2001 study American college students were presented with essays purportedly written by students at another local university. The essays were entitled: 'The most important things I have learned about human nature'. One essay argued that '. . . the boundary between humans and animals is not as great as most people think'. It went on: 'Although we like to think that we are special and unique, our bodies work in pretty much the same way as the bodies of all other animals. Whether you're talking about lizards, cows, horses, insects, or humans, we're all made up of the same basic biological products.'

An alternative essay claimed, 'Although we humans have some things in common with other animals, human beings are truly unique.' It continued, 'Humans have language and culture. We create works of art, music, and literature that enable us to live in an abstract world of the imagination – something no other animal is capable of.' In the study, the researchers found that when provoked to think on their mortality, college students showed a clear preference for the essay emphasising human uniqueness compared to the essay on human creatureliness.

It has since been found that if people are primed to think about death, they are more likely to give to an animal charity that highlights the distinction between humans and animals. In another study on attitudes to companion animals, psychologist

Ruth Beatson asked volunteers to read a short paragraph on humans and other animals. When prompted to think of mortality, participants evaluated their pets less positively. The findings are especially noteworthy as these were pet owners with generous attitudes towards other species. It seems our fears and our beliefs about being animal are locked in a strange embrace.

Though he is out of fashion now, Freud offered up a vision of humans as beings that had repressed their animal instincts at the expense of their sanity. Freud's great insight was that a huge territory of anxiety underlay common human experience. Some of his conclusions were downright ridiculous, but he nonetheless raised the spectre of an animal howling at its image in the mirror, a thinking creature disturbed by a split in its perceptions and impulses. It is an image that many have wrestled with since.

Ernest Becker was a cultural anthropologist who wrote a popular book on the tendency for humans to seek ways to overcome their existential anxieties. Humankind, he wrote, 'is out of nature and hopelessly in it . . .', trapped in 'a material fleshy casing that is alien . . . in many ways – the strangest and most repugnant way being that it aches and bleeds and will decay and die'. The body is a 'hurdle for man', a 'threat to one's existence as a self-perpetuating creature'. Becker was not alone. He was in conversation with a group of theorists in the early to mid-twentieth century, among them Rollo May, Erich Fromm and Otto Rank. May, an existential philosopher whose brush with death and three years in a sanatorium were fundamental to his theories, considered anxiety to be the prevailing condition of humans. As he saw it, we all do mental battle with our animal being and, underneath this, with an amorphous fear of 'nothingness'. And according to Becker, Fromm believed that what lies at the heart of human experience is a *paradoxical* nature, the fact that he is half animal and half symbolic'. For Fromm, this mental tension gives rise to the compulsion to cut the animal half loose.

I have a second-hand copy of the 1973 work *The Denial of Death*, for which Becker became known. The book is highlighted throughout and contains the comments and thoughts of the previous owner. Whoever had passed this copy to a bookshop to be resold had been struggling to come to terms with the events that surrounded the terrorist attacks on the World Trade Center. Inside, on the first page, is a large note: *Q: Who were the 'true' heroes of 9/11?* In the preface, the previous owner drew an emphatic arrow on one section of the page. The passage reads as follows: 'The idea of death, the fear of it, haunts the human animal like nothing else; it is a mainspring of human activity – activity designed largely to avoid the fatality of death, to overcome it by denying in some way that it is the final destiny for man.' Where Freud thought neuroses emerged because people didn't really believe they would die, Becker saw the whole of society as a codified hero system. 'Society everywhere is a living myth of the significance of human life,' he wrote.

In 2010 campaign designers Fred & Farid created a series of adverts for Wrangler jeans with the title 'We are animals'. 'Man is an animal,' the designers said, 'but he no longer knows it.' A short while later, some research was conducted on the ad campaign by American psychologists Sheldon Solomon, Tom Pyszczynski and Jeff Greenberg. Using the ideas of Ernest Becker, Solomon and his colleagues hypothesised that subconscious fears of death generate feelings of meaninglessness and insecurity about our place in the world. When fears are high, people favour the idea of human uniqueness as a way of protecting us. In the study, an alternative advert was produced with the line, 'We are not animals.' Those prompted to think about death preferred the doctored advert that presented human uniqueness.

It isn't difficult to find cultures that use meaning systems to manage existential fears. Spirit guides into the afterlife exist in most cultures. From the ancient fertility cults of Osiris, the germinating seeds of wheat sown in the shape of the pharaoh, as if by each green frond he rises from the dead, to the reincarnating Ātman (soul) of Hinduism, belief systems are unified in lifting us out of danger. Of course, many worldviews see death as a transition and a normal part of life's rhythm. But that doesn't necessarily mean that individuals don't continue to buffer themselves from their mortality.

In Varanasi, with the Ganges river running through it, death and daily life mingle. In one of the holiest cities in India, people enter the waters there to refresh their souls and to feed themselves. Above the waters is a haze of human particles, the ashes of the hundreds of bodies cremated by the shores each day. During monsoon season the corpses can come loose from their mooring stones and float down the river. Death and the river flow like weird twins.

In the Hindu worldview, people are endlessly reborn. The completion of each cycle is achieved through *moksha*, the cremation of one's flesh, when the soul breaks the bonds of an animal life and returns to the pure spirit of Brahman. In one study from 2010, researchers worked with two sets of people living near Varanasi: those exposed to death regularly and those with low exposure. What they found was that those individuals, such as farmers, working at a distance from the Ganges, experienced a marked increase in their defence of their worldview when encouraged to think of death. The same spike didn't take place among those working as funerary attendants, but only because they already showed a chronically high impulse to protect their cultural worldview. The researchers concluded that exposure to death didn't habituate people to dying but to defensiveness.

Similar studies were reported among Aboriginal Australians a decade or so earlier. This is an especially compelling case, as the cultures of Aboriginal Australians involve a deep attachment to the rest of the living world. Provoking thoughts of death here appeared to cause individuals with exposure to both mainstream Australian culture and their indigenous identity to strengthen their belief in their Aboriginal value systems.

But death isn't the only flagrant evidence of our animal reality. There's also the small matter of how we're born. Women, because of their visceral role in reproduction, serve as potent reminders that we are animals. After repeated studies on attitudes towards women, psychologist Christina Roylance has concluded that believing humans to be a unique and superior species might offer the benefit of existential confidence, but such beliefs may come at the cost of derogating women. Most often, suppression follows male control of female reproduction. But if female bodies pose a threat to the ideologies that justify patriarchy, the doubling-down may be that much greater. In other studies, Jamie Goldenberg found elevated reactions against public breastfeeding among those primed with a threat. And Mark Landau's experiments have shown that heterosexual men experiencing higher levels of existential fear find overtly sexual women less attractive.

In making sense of our times, then, it's worth remembering that – consciously and subconsciously – many people don't like to be reminded that we are animals, so much so that we may unconsciously rebel against anything that seems to place us in this category. Excreting, menstruating, all the leaky and messy parts of our organic being, can produce an almost imperceptible but significant shiver in our minds. When added to those things that disturb us – injured bodies, diseased or dying bits of us – we find that being an animal pricks our imaginations, especially when we feel vulnerable. Even a quick glance at a damaged body prods an organ of horror that we can neither see nor control. What is conjured – like an anti-genie who steals rather than gives – is the vision of our decay into nothing.

On the one hand, the idea that we are frightened of our deaths seems unremarkable. Yet it supports the idea that we don't think we're an animal because we don't want to be one. 'What joy can there be in life,' Cicero lamented, long centuries ago, 'if day and night we are forced to consider the inevitable approach of death?' The solution, Cicero writes in *On Living and Dying Well*, is the possibility of an immortal soul. 'This is how I want things to be,' he writes. 'And even if they aren't, I still want to be convinced that they are.' Cicero understood that our fears can be sweetened by the possibility of uniqueness.

These days, we like to think of our times as being largely free of many of the direct dangers that other animals face. We've reduced most of the populations of fierce mammals. We've made our homes stronger and more sanitary. We have a raft of medical technologies to extend and safeguard our lives. Even our wars with one another have become less lethal in the past decades, although the potential remains. The unusual pattern of the past few centuries has left more of us than ever assured that our bodies will be fed and given the opportunity of a long life. This has generated a justifiable feeling of control. But the less of ourselves we surrender to natural forces, the more unwilling we become to surrender anything at all. So it is that many of us are less prepared than ever to see ourselves as animals.

Antidotes for extinction

We are living through an age of extinction. A study on birds published in 2019, led by biologist Melanie Monroe and BirdLife International, looked at species losses from a range of different perspectives. Comparing background rates with extinctions from the past five hundred years, they calculated a figure of nearly two hundred lost bird species rather than the four or five they might have expected. Looking at categories from least to critically endangered, they found that the general trend was for more species of birds to become endangered. And looking at the expected lifespan of a species in deep history, they believe that vertebrates generally exist for three million years before becoming extinct. Using current rates of extinction, a species is lucky to make it past five thousand years. Similar stories can be told across many of the world's taxa.

I began studying extinction in my twenties. Then, it was a struggle to draw much attention to the subject. But these days extinction is in the daily news cycle, with constant headlines about the potential dangers we and other animals face. This may not sit well with human psychology. What might extinction prompt in the mind of an animal that not only grasps the significance of such an end, but knows it is the cause? Extinction is a challenge to any society that bases its own moral status on being the peak of an evolutionary pyramid. And so, while we have conservation efforts under way around the world, other solutions are being sought with equal passion.

In the closing scene of H. G. Wells's 1936 futuristic film *Things to Come*, Oswald Cabal, the governor of a technocratic society, watches a capsule elide into the night sky. Shot into space from a giant gun, the capsule carries Cabal's daughter and the son of his friend, Raymond Passworthy, and a host of other young interstellar migrants. Gazing on as their children leave the Earth, Cabal argues that humanity must always go onwards, overcoming

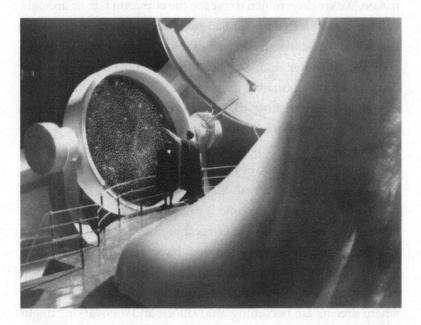

the limits of nature, the straps of the planet, and finally 'all the laws of mind and matter that restrain him'. But we're just little animals, Passworthy interjects. 'Little animals,' Cabal repeats.

'And if we're no more than animals, we must snatch each little scrap of happiness, and live and suffer and pass, mattering no more than all the other animals do, or have done.' But Cabal doesn't believe what he's saying. On the contrary, he is convinced that humans can escape the conditions reserved for all other Earthborn creatures. As he tells it, if we don't want to simply live, struggle and die, we have no choice but to strive out into the immensity of space. It is all the universe or nothingness. 'Which shall it be?' Cabal asks. 'Which shall it be?'

Passworthy longs for a future in which we are exceptional. The root meaning of exception comes from the Latin word *excipere*, which means 'to take out'. In thinking of ourselves as exceptional creatures, we are taking ourselves out of the rest of nature. We are determined to escape the common fate of animals. What is certain is that the human story has always been a struggle against an endlessly shifting threshold. It is like the rainbow's end. We run after and, by each onward movement, alter the place in which a potential limit exists. We have pushed out this threshold so often now that we have come to believe it's an illusion. So what is fascinating about our times is that a large number of us are convinced that there is no end point for humanity, no boundary beyond which our species will cease to exist and become a part of the Earth's history.

Now that we have become a prime agent of extinction, we have begun to fantasise about escaping our own. Many of the world's largest tech corporations have recently invested in robotics, artificial intelligence, interplanetary colonisation programmes and brain enhancement. We live in a strange era, where dreams are becoming innovations and innovations might abolish what it means to dream in the first place. We can observe, too, that avoiding extinction has become the favourite justification for many of those who seek investment in such schemes.

One solution for extinction is to go into space. At present, there are powerful elements in our societies that seek to expand into the solar system. The Mars Society was founded in the late 1990s to educate and lobby for the human settlement of Mars. Not so long ago, I went to visit one of the founders, Robert Zubrin, at the Mars Desert Research Station in Utah, a collection of white space-age buildings in an arid landscape of dust and dinosaur bones. Mars is a resource, he told me, because 'resources are created by human imagination and technologies'. He points to the progress already achieved in the past few hundred years in terms of life expectancy, wellbeing and productivity.

Zubrin and others are part of a backlash against the Club of Rome's *The Limits to Growth*, a 1972 report on a computer simulation of economic and population growth against a background of finite resources. The report's authors had suggested that, without changes, the obstacles to human expansion on Earth would kick in at around 2072, leading to a sudden and uncontrollable decline in population and industry. It was roundly criticised. But in recent years some of the predictions of the model have been shown to be accurate. Correct or not, *The Limits to Growth* motivated some people to prove it wrong. Zubrin was one of those. He remains genuinely concerned about the pessimism that has entered into the public debate and has an infectious confidence in human inventiveness and the need for an optimistic outlook.

But not everybody thinks such ambitions are realistic. A study by Bruce Jakosky, who was principal investigator of the NASA-funded Mars research project, and geologist Christopher Edwards argued that there's insufficient carbon dioxide to allow humans to create an atmosphere on the planet. Even more problematic, Mars may once have had a molten core and magnetic field like the Earth but not any more. Without this, it's almost impossible to create a thick enough atmosphere for animals like us.

As a NASA veteran and a scientist with a lifetime researching the origins of life, Michael Russell thinks 'we've a better chance at solving poverty than making a mirror Earth somewhere else'. He sees the obsession with Mars as a refusal to admit that 'you can't beat thermodynamics'. The Roman Empire hit the skids because it ran out of forests. The Industrial Revolution was made possible by greater free energy close to the centres of industrialisation, but our energy sources have turned into a problem for today's societies. Modern humans are determined to find a new means of generating more energy for themselves. There's little doubt we will achieve this. But it always comes at a price.

Maybe later this century we will have one-way trips to Mars supported by Earth. But, even if possible, it would take a monstrous amount of money to build an atmosphere. And we might need technologies like DNA-repairing treatments to survive the radiation on the surface. We might argue that there's nothing essentially wrong with the ambition to go out into the solar system – we might see it as an exhilarating idea. But there can be something wrong with the reasons to do it. At present, the point of departure to Mars is a vanguardist one funded by billionaires. On Earth, water is the key resource. What if we had companies that wanted communities to pay for air on Mars? Is there anything in how finances are currently managed on our own planet that would lead us to presume there would be good funding principles on Mars?

Rob Lillis is a Mars specialist and a NASA-funded researcher at the University of California at Berkeley. He is broadly behind the idea of getting to Mars but worries if the people who most want to go are part of a libertarian, anti-regulation culture. There's no obvious sign that those trying to build new societies elsewhere have considered overhauling our current ethical structures. 'If they had,' he told me, 'wouldn't they invest first in improving the inequities of individuals' lives on Earth? Why aren't they putting their money and inventiveness into getting plastics out of the oceans or innovating new, clean materials? Why the obsession instead with immortality and colonisation?'

But it isn't only Earth that we now seek to escape. In other ways, our era is revealing itself in the ambition to escape our own bodies. When Yeats wrote of his heart as 'sick with desire/ And fastened to a dying animal', in his poem 'Sailing to Byzantium', he gave voice to an almost universal feeling. But his plea, 'gather me/Into the artifice of eternity', has been taken seriously by the technologists of the early twenty-first century. In a world of uneven riches, large sums of money are changing hands in the pursuit of eternity.

Some have been seeking biological immortality through an enzyme called telomerase, which repairs telomeres, the bits at the end of DNA that shorten as we age. Others look to electronic immortality, the fusing of our conscious experience with machines. Spike Jonze's 2013 film *Her* depicts the romance between an intelligent computer program, Samantha, and a young human. 'I used to be so worried about not having a body,' Samantha says, 'but now I . . . I truly love it . . . I mean, I'm not limited. I can be anywhere and everywhere simultaneously. I'm not tethered to time and space in a way that I would be if I was stuck in a body that's inevitably gonna die.'

Unconcerned by what this might mean for the wider balance of life on our planet, the research progresses. The first glimmer of success was a pill called RTB101, the invention of Boston biotech company resTORbio, which appeared to reduce the effects of ageing on our immune systems, improving resistance to viruses like the common cold and influenza. RTB101 came out of a huge investment in initiatives to fight against the limits of our lifespans. Recent studies from a team in California suggest that it might be possible to reverse the body's epigenetic clock, which researchers hope can reduce a person's biological age. Using a mix of drugs, the trial found reductions in the interpreted age of the immune system and age markers on the genome of the participants.

Commercial start-ups that turn fear of death into a lucrative product have cropped up around the world, from Rejuvenate Bio at Harvard, which seeks to move from extending the lifespans of treasured pet dogs to allowing humans to live into the hundreds with 'the body of a twenty-two-year-old', to the Indian company Advancells, which aims to use the 'limitless potential of Human Stem Cells in providing a natural cure for any ailment that our bodies suffer from'. Nobody doubts that such companies believe it is a straightforward benefit to improve the lifespans and health of ageing humans, but at what cost? The exploitation of our biology is seen as the new frontier of progress and salvation. But, in effect, we have to kill the idea of ourselves in order to undertake the research that will save us. These are questions not only about what kind of life is worth living but what kind of way it's worth dying.

Google, the world's largest internet company, is used by billions of individuals worldwide, searching through its platform more than a trillion times a year. In its capacity to influence, the culture of Google should interest us. It matters if a senior engineer at Google has made it abundantly clear that his strategy is 'not to die'. 'Death is a great tragedy,' Ray Kurzweil has said, 'a profound loss . . . I don't accept it . . . I think people are kidding themselves when they say they are comfortable with death.' If he can't solve the problem of mortality in his lifetime, his body will hang in liquid nitrogen in the Alcor Life Extension Foundation in the dusts of Arizona until someone else has figured it out. He's not alone. There are hundreds of bodies, some just heads or brains, in storage cocoons, awaiting their moment of restoration. In KrioRus, on the outskirts of Moscow, on the flats of Michigan at the Cryonics Institute, humans and their pets keep their decay on hold. In journalist Mark O'Connell's study of transhumanists, *To Be a Machine*, he quotes Natasha Vita-More, a prominent member of the movement, as saying, 'We're a neurotic species – because of our mortality, because death is always breathing down our necks.'

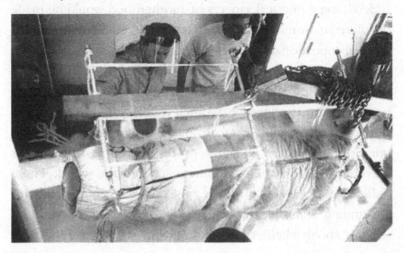

Meanwhile, as we wait on the chance to live forever, the alternative is to upgrade our biology somehow. Modern eugenicists seek to create a better life for the living. They rarely broadcast the fact that we'd have to stop having sex and gestating as we do at present. Commonly, female babies have a million or more eggs already inside them. It's a little like a matryoshka doll, each generation nested infinitely inside the other. Only in recent years has it been discovered that women's bodies may grow new eggs during their lifetime. There are some who would prefer to pluck these eggs and fertilise them outside of the womb, so that superior embryos can be selected for. Advocates argue that humans have always interfered with sexual outcomes, whether by choosing a mate or using a contraception. But this promise of continuity is misleading. Mate selection and condoms are matters of energy, not value. None are equivalents to deciding what we think an improved child consists of.

What those beguiled by enhancement dreams refuse to acknowledge are the bodies that are left behind. The flesh of the poor, the different or the disabled, the flesh of mothers. Indeed, some physical aspects of motherhood would be made obsolescent. Some may see this as a boon. In recognition that the disproportionate burden on mothers has restricted what individual women can achieve in their lifetimes, motherhood has become something to be limited for the sake of women's progress. This is nonsense. Women gain through freedom and respect, not by saving them from being mothers. We should question instead the endgame of those who wish to master everything, even their own emergence into the world. These radical monogenesists hope to be the gods of their own nativity.

Running alongside the fancies of genetic engineering are debates about whether non-living forms of engineering such as computers and machines will enhance or even replace humans. These will be immortal analogues of the personal essence of who we are. From an animal that came to know itself, we will become a knowledge that came to abandon itself. But will these ethereal selves age? Will they live through the imaginative outlook of a child? When individuals speak of uploading our skills and personalities, they are concerned only with the spoken subjective consciousness that will emerge later in a human life. The consciousness of a human embryo doesn't enter the picture. Human embryos have no knowledge of a 'you'. They flip and grow on the measures of blood and hyle. Does that make the experience a needless phase?

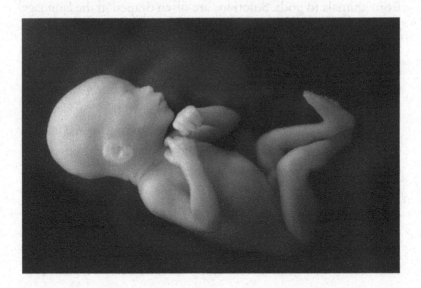

Of note is the fact that many who've invested in these kinds of big-tech solutions to being animal have also been involved in the analysis of our own possible extinction. At each of the world's leading universities can be found recently inaugurated centres of extinction risk analysis, one at Cambridge, one at

Oxford, and the Future of Life group at MIT. The foci of these studies include the malicious misuse of new technologies and the succession of natural human populations by engineered superhumans. It is worth considering, then, whether these sorts of studies come with their own risks. Whether the danger is in our over-utilisation of our planet's resources or in new inventions like artificial intelligence, the justification for each is human supremacy. But if being confronted by potential existential dangers can cause us to overemphasise human exceptionalism, studies on existential risk may exacerbate, even accelerate the risks we face. The extinction of our own species is the ultimate denial of the idea that humans are on some kind of trajectory from animals to gods. Solutions are often draped in the language of salvation and progress, rather than humility or compassion. It's notable that donors to these kinds of existential risk projects include some of the biggest figures in technology. The possibility that human progress as currently imagined is but a myth may be of particular concern to those who believe they've ascended through the ranks. Perhaps we shouldn't be surprised to see dreams of transcendence arise among the wealthy entrepreneurs of some of our wealthiest countries. It may seem from such a viewpoint that the only choice left is the complete removal of humans from all earthly limitations. The alternative – to climb down and to share, to conceive of limits and seek progress in ideas rather than materials – may cut across the grain of nature.

A better alliance?

We may be a defensive, territorial animal capable of appalling behaviour. We may be peculiarly ill at ease with the state of being animal. Still, it's important to remember that this is only a fraction of what we are. A large part of us seeks to be thoughtful, compassionate and friendly. The same studies that have shown how prickly and hierarchical we can be have also shed light on some of our more endearing qualities. The research that exposed the antipathetic responses that are possible towards sexual partners also found that focusing on love offers an abstract representation that can soften this. In some of the studies on people with extreme intergroup hostility, first contemplating compassionate values reduced their hostility towards those with opposing views.

A group of researchers in Canada have undertaken extensive studies into dehumanisation. For Kimberly Costello, the tendency to view some people as more like animals 'may be rooted in basic hierarchical beliefs regarding human superiority over animals'. But rather than assuming that human superiority is a fixed event, she was curious to find out if narrowing the perceived gap between humans and other animals changed the dehumanising urges. She found that this way of thinking is surprisingly suggestible.

In experiments designed by Costello and her colleague Graham Hodson, university students convinced of strong human uniqueness were generally more prejudiced towards immigrants, while those who subscribed to the 'animals are similar' category exhibited the least immigrant dehumanisation and higher empathy. Strikingly, Costello and Hodson discovered that it was possible to reduce or temporarily eliminate someone's dehumanising beliefs about other people not by manipulating their insights into their fellow humans but by prompting these individuals to reconsider other animals. Such a reappraisal failed if someone was asked to drag humans down to the level of animals. But it worked if someone was called on to see both humans and other animals as precious entities sharing valuable needs and feelings. In another experiment, after watching videos that encouraged insight into the lives of other creatures, children's belief in a strong separation also decreased. In both cases, the idea of a better kind of relationship buffers against a less pleasant impulse. But this isn't because the idea itself is so convincing that our rational self prevails. It's our bodies that do the work. Physical processes support us in forming affiliations with each other. Ideas trigger the change from defensiveness, but our bodies bring it about.

In recent years, scientists have tried to home in on the physical events that result in the phenomenon we've come to call empathy. The consensus is that empathic behaviours arise from a complex of processes and experiences without a single cause. Some work has focused on the ability we and some other mammals have to share emotional states through mirroring. Mirror neurons, first isolated in macaques, fire in the brain when doing something like grasping but also when observing others doing it. The theory is that we and other animals directly map representations of movement or sensation of observed action. This may come in many guises. If we see a picture of someone with a look of disgust on their face, for example, we show similar brain activity to the direct experience of something disgusting. But so, too, when we see an act of love, we smile in recognition. This faculty, helpful in everything from social learning to social navigation, may be a crucial instrument of empathy. Other research focuses on mimicry. Many animals observe others and synchronise with them. But, for humans, we need only imagine someone else's experience for there to occur an outbreak of cortisol or a sudden burst of laughter. In this way, an idea that causes us to mirror a positive emotion can dramatically alter the way we respond.

What the mechanisms of empathy achieve is a temporary insight into the experiences of another human or even another species. But what we call empathy doesn't necessarily end in a response that is commensurate with what our empathising allows us to see. What we do with the insight is then modulated by the relationship we want. It's more than likely that we have various processes to defend against the automatic mimicry-like processing if we don't want to identify with another individual for whatever reason.

Across the millions of species that live on this planet, many kinds of relationships are possible. In the scientific literature, those who study animals write of several ways in which different species come together to forge partnerships of sorts. Commensalism is the term for an alliance of tolerance. Terms like this should be understood as useful but imperfect ways of talking about things we don't fully grasp. What the odd word commensalism tries to capture is the phenomenon where one animal gains one or several benefits from proximity to another, while the other tolerates their presence because, while they don't necessarily gain, they don't lose out either. An example of this might be the many African birds that use large mammals as perches or for safety from predators.

In other cases, a relationship is or becomes what is called mutualistic. Among those birds that get carried about on the backs of massive animals are oxpeckers. They gravitate to mammals with a considerable infestation of ticks. In this way, they not only gain a source of food but they benefit the other animal by reducing the number of parasites on its body. Zebras and ostriches, both herd animals, also forge two-way alliances. Zebras have fantastic hearing and a good sense of smell but poor eyesight. Ostriches, meanwhile, are sharp-eyed but less sensitive to the scent of danger. As a consequence, zebras and ostriches, both preyed on by the same predators, travel together; one acting as the eyes and the other as the ears and nose. If trouble comes, one warns the other.

All kinds of biochemistry and cognition can be involved in the relationships that emerge between animals. But what we've come to call bonding is something slightly different again. Used

first to describe the coordinated selective reproductive behaviours of animals like seabirds, and later the behaviours between parents and offspring, bonding has become a more general-purpose term to explain those mutual relationships that go beyond temporary convenience. When we stroke a horse, the animal's heart rate slows and oxytocin concentrations increase. When a human interacts with a familiar dog, they undergo a decline in blood pressure greater even than being at rest.

Influenced by oxytocin, which is produced in the hypothalamus of all mammals, and by other organic chemicals such

as dopamine, a bond between two people or even two species is always mutual. In bonds of any kind, something good is shared intimately, shot through with pleasure. Even a more distant but mutual relationship involves a shared good. A pair-bond between two seabirds may be more instinctual. But, either way, the relationship is only rewarding for each individual because of a shared advantage. Whether we choose to admit it or not, much of what we're really talking about when we speak of kindness, even of happiness, originates in these kinds of relationships.

But relationships can follow other pathways. Think of the house mouse, *Mus musculus*. It would appear that mice adapted to the sedentary lifestyle of humans during the Late Pleistocene. At this time, the ancestors of those little black-eyed beings that now nest in our attics and cellars became domesticated because of the greater success of those mice that could tolerate human presence. But the relationship between mice and humans didn't turn into a bond. In truth, mice either had no positive impact on human lives or were a pest. The advantage flowed in their direction. But when watching a dog and a human playing together on a beach, we can recall that this may have been made possible because some wolves in the ancient past stood to gain by foraging from the leftovers at human settlements. And, somewhere along the line, what these individual animals gained by expanding their relationship with humans turned into a bond where both species took something of value.

Natural selection against a background of human presence went yet another way for some ungulates. One theory is that domestication followed from those individuals that were sufficiently placid or bold to begin grazing in the wheat and barley fields of the Fertile Crescent. What suited them for a while gave way to a relationship that is now, most often, only to the advantage of the animal that planted the wheat. Today, few goats or pigs or cows have a relationship to humans that is much good for them.

What seems incredible about life on Earth is the emergence of something now widely referred to as an ecosystem. In between the two global wars, an English botanist who once spent a year studying with Freud in Vienna gave us the concept of the ecosystem. 'Though the organisms may claim our prime interest,' Arthur Tansley wrote in 1935, 'when we are trying to think fundamentally, we cannot separate them from their special environments, with which they form one physical system.' Again, we're only ever doing our best to find a word to explain something that is extraordinarily messy. None of us exactly knows what we mean by 'ecosystem'. Still less how to think of our own role inside one. But once there is a plentiful variety of life forms, it seems, the world arrives at a temporary condition that acts to enable life.

We were once a fully integrated part of African ecosystems, the activities, lifetimes and populations of our genetic ancestors directly controlled by predators and scarcity of food. But, as anthropologists like Ian Tattersall have pointed out, somewhere across the history of our lineage, cultural and genetic traits allowed us to evade some of the limits placed on us by large predators and we began to find collective means of defence and survival. Once these emerged, the impact of our species started to alter. A paper published in 2018 argued that *Homo sapiens*, and possibly other human species, began causing the extinction of large mammals at least 125,000 years ago. At first, such losses took place in Africa, but the pattern began to spread into other regions as humans migrated.

Large mammals, whether predators or herbivores, have wide-ranging effects on ecosystems, from the recycling of nutrients to limiting populations of other animals. Under the effects of our species, other mammals have been slowly getting smaller because we've caused the extinction of larger species and individuals. 'Ecosystems are going to be very, very different in the future. The last time mammal communities looked like that and had a mean body size that small was after the extinction of the dinosaurs,' said study author Kate Lyons. And our ancestors had other impacts as they migrated. Most often our presence was followed by degradation of vegetation cover along with agricultural erosion and pollution. In 2019, it was announced by Robert Watson, chair of the Intergovernmental Science-Policy Platform, that 'the health of ecosystems on which we and all other species depend is deteriorating more rapidly than ever'.

We don't need to get too hung up on whether something is natural or whether our understanding of something in biology is complete. If we look at the way *Homo sapiens* has interacted with the rest of the living world, what we can see is the emergence of an unbalanced relationship with the Earth and its other species. The goods have flooded in our direction. That's pretty straightforward. And there are consequences to such an imbalance. Many of the dilemmas we now face have arisen from this. But if we do nothing, there's no doubt that something external will put the brakes on us. Surely, then, it would be better to contrive our own limits? For this, we will have to play to the strengths of our social psychology.

I think our ancient hunter-gatherer ancestors may have been on to something in seeing agency in everything. Perhaps for them it was purely a practical means of out-thinking their prey that sank into all the reassuring symbolism of their worldview. But what unites the different routes to the human experience of empathy is the willingness to think about someone or something else and match our behaviour to what the other needs. If we want something more like a mutualistic approach to the life around us, we will have to see into the needs of other organisms. We are still only at the beginning of our discovery that all animals are profoundly intelligent and complex. Yet everything we've discovered about our planet's life over the past few thousand years gives us a rational basis for reconsidering the value we give to other living things.

One thing we may have to rethink is death itself. An ecosystem includes death as a normal and necessary feature. Perhaps we know deep inside us that death is not just ordinary but essential. It isn't to be sought. But neither is it to be feared as a strike against who we are. A refreshing perspective on death was offered recently by writer and biologist Bernd Heinrich. For Heinrich, the whole Earth is an organism, ceaselessly releasing and regathering energy. Death, by such measures, is a gift that can be traded to bring into the world more bees and butterflies, more humans and hawks, more blossoms and wheat. Heinrich sees a practicality in all this. In resisting this exchange, humans have turned death into a monster and being animal into a trap. Yet 'all bodies are built of carbons linked together, later to be disassembled and released as carbon dioxide . . . The carbon building blocks that make a daisy or a tree come from millions of sources: a decaying elephant in Africa a week ago, an extinct cycad of the Carboniferous age, an Arctic poppy returning to the earth a month ago.' It's only humans who don't want to join the party.

THE JOURNEY-WORK
OF THE STARS

I believe a leaf of grass is no less than the journey-work of the stars,
And the pismire is equally perfect, and a grain of sand, and the egg
of the wren.

Walt Whitman

We are stardust

The darkness of the space through which we spin is not a dead or an empty colour. It's a living black. It is the shadow through which the quintessence of us and all other living things on Earth once journeyed. Some-where out there, every second or so, a star dies in extravagant style, spitting out the elements of life in

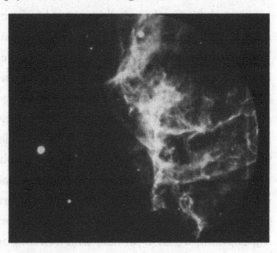

its final breath. Its death is a titanic explosion in which more energy is released in one go than our sun will produce in the full course of its life. It's a hell of a way to go.

But in a matter of weeks, the huge star becomes a shimmering graveyard of substance in the quiet of the cosmos. Out of this final and permanent quiet we can see what it gave us, a great outpouring of elements, sulphur, argon, cobalt and many more, forged by the massive swell of energy and neutrons. We came to be from the death of a star. In time, our own star will die too. We fashion our convictions from the dark exuberance of a universe in which our lives are tantalisingly brief. The sun, life-giver, for so long a symbol of magic and holiness, is but an ageing star whose helium heart will one day fail, absolving itself of its energies, bulging into a red colossus, slowly consuming Mercury, Venus, then our Earth, finally lacing its grave in blue and mauve and opal remnants of itself. Elsewhere in the universe, other, larger stars might break apart violently and suddenly, shattering in an instant of brilliant light, spewing out the very elements that once gave life to us, beckoning them, perhaps, into the chance emergence of another marvellous being.

We and the Earth and all its other life forms are made of the shrapnel of a dead star. By the time we reach adulthood, our bodies hold within them more than sixty elements, the residue of this primal kind of extinction. We are a sheaf of empty space and ancient electricity, an unimaginable quantity of atoms, carrying protons and neutrons and somersaulting electrons. Our bodies are the sum of many trillions of cells, a proportion of which renew themselves multiple times throughout our life. We are the billions of letters of a genome and a whole fizzing microbial community of bacteria, yeasts, viruses and helminth parasites such as roundworms and tapeworms. Assuming no impediment, we are a primate with a brain possessing vast quantities of nerve cells linked together into a large complex, out of which, somehow, our thought processes emerge.

Everything we see out there in the woods and deserts and seas is a relative. A cousin. A second cousin. Or a second cousin a thousand times removed. But they are kin, nonetheless. Most people think of worms as little more than a slightly yucky creature in the backyard or as something to be dug up to bait fish. But the first evidence for bilateral symmetry that gives us our slight hourglass shape is found in a species of worm that lived six hundred million years ago. The first appearance of a backbone, the source of the shivers down our spines, is in a jawless fish that lived alongside the trilobites.

Our muscles contract and move like those of other animals because muscle fibres, responding to stimulation from nerves, are altered by chemical energy from the breakdown of food molecules captured in an energy-carrying molecule we call adenosine triphosphate found in the cells of all living things. Like all other life on Earth, we are the offspring of shared processes, a confederacy of matter that can metamorphose into the seemingly limitless wonders of life. All the while, assuming we progress normally from ovum to pensioner, humans possess a psychology that is sensible of its condition. All this amounts to us being a thing that fascinates and frightens us. We are an animal.

On the one hand it's obvious to most of us that we are animals. When asked, people are convinced this must be true. Yet we're still told society must be approached as if we're not. For millennia we've viewed ourselves as separate from all other creatures. The thought that we belong to the same physical world as the rest of life on Earth is one that successive generations of people have found impossible to accept.

The Maasai are the chosen people of the demigod Enkai, who blesses them with the cows he passes down through a kind of cosmic rip. In some Chinese myths, a goddess named Nü Wa becomes lonely after the death of the cosmic being P'an Ku. She gathers mud from the edges of a pond and forms it into a human. She likes the little dancing men and women so much that she peoples the world with them. Thanks to the poet Hesiod, we have been handed down ancient stories of the emergence of the volatile and jealous gods. Out of this flawed universe humans arrive, stubbed from clay. But Prometheus gifts us the ability to stand upright and command fire.

These myths may have been in Rudyard Kipling's mind when he portrays the village children in *The Jungle Book* moulding people from mud and putting reeds into their hands to pretend 'that they are gods to be worshipped'. At the same time, the Bandar-log, the monkeys of the jungle, are desperate to be humans. They hide away in the ruins of a once-grand temple. 'We are great,' they coo. 'We are free. We are wonderful. We are the most wonderful people in all the jungle! We all say so, and so it must be true.'

What myths do so well is to give us a condensed reality. Instead of thrashing out ideas and wasting time and energy on the nitty-gritty, myths offer a simple symbolic shorthand for who we are and our relationship to the world. In the twenty-first century the myths of modern societies tell us we're not animals. Our enormous skills give us unique value among all other life and justify the harms we do. We don't cause injury or destruction because we're animals but because we have reason. We flourish at the expense of the rest of life because flourishing is a human right.

Of course, there've always been thinkers who've rebelled. In the older traditions of humans who lived among wild animals, there was usually the belief in some special phantasmal aspect of us. Yet humans weren't alone in possessing essences of this kind. All of the living world was bound by spiritual force. Even at the onset of the Western belief in human uniqueness, Aristotle's mysterious student Dicaearchus claims in his Corinthian discourse that there is no such thing as the soul. The word 'animal' should also be used with caution, he says, as there's no animating force, no separate, soul-like property in living creatures. The sensations and experiences of life drench animal bodies at every level. They can't be cored from the flesh like the pit of a peach. But his view – in many ways the most interesting – has proved the least popular.

The central message today is that we are special and enjoy special rights to exercise our animal appetites. Other animals hunt because they must, but we humans make use of other animals because we are more important than them. To make sense of this, we have thought of ourselves as having a human and an animal half. In this worldview, the human half saves us, whether this comes in the form of a soul or in a rational sense that gives us both freedom and the prerogative of our species. So it is that for at least several thousand years now humans have been alone in their quality.

But there's also an irrational kernel inside us, a fear of being animal that was the price of becoming a person. As we became self-conscious, our personal view floodlit what can be a danger to our bodies and laid bare the inescapable danger of death. We've become the conundrum of an animal that doesn't want an animal's body. What was survival has re-emerged, by a long, curious path, as psychological imperative. Other animals don't have to justify themselves to themselves. But humans seek what might give their lives a meaning that no other animal possesses. If we don't belong to the rest of nature, its dangers can't reach us.

But rather than reassure us, this strategy has left us reliant on a falsehood. The myth of human exceptionalism is as unsettling as it is irrational. The idea of our superiority runs with mercenary and sometimes aggressive features of our psychology. Being animal is a kind of syndrome for us, a peculiar combination of symptoms, emotions and opinions. It's something we deny, manipulate as a weapon and seek to escape. At other times, being animal is given as a reason for our actions and, as often, the excuse for them. And so our lives are spent quietly haunted by the truth of a connection to nature we can barely admit. In our dream of uniqueness, we forever dance on the hot coals of a landscape that disquiets us.

Eugene Thacker, a philosopher based in New York, writes of the 'precipice of philosophy . . . [an] abyss, where there is neither certitude nor knowledge, nor even a single thought'. Descartes, he tells us, peered over this edge and didn't like what he saw. In his famous dictum, *cogito ergo sum*, Descartes says 'he will never bring it about that I am nothing as long as I think I am something'. 'From this,' Thacker writes, 'flows an entire legacy of philosophical thinking . . . and the privileging of human consciousness over all other forms of being.' Even among the staunch humanist philosophers like Kant, there's an undercurrent of fear. Only humans can reason, he announced, therefore only human life has reason. But in his lesser-known essay 'The End of All Things', Kant seems to falter. 'The rational estimation of life's value,' he admits, may not point to the primacy of humans.

It has always been a dubious solution to build the worth of our lives on some absolute dividing line between us and the rest of the living world. But it didn't present such difficulties in previous centuries. When Thomas Paine sketched his *Rights of Man*, we didn't yet have the theories of evolution. When Pico della Mirandola imagined a God that had placed man at the centre of the universe by dint of reason, we were nearly four hundred years away from the discovery of DNA. But we don't have the luxury of ignorance any longer. We are living in a time when we have to rethink everything around us. Science has shown us that we're an animal, slowly shaped from ancestral forms. Before 1859, we were a gift of the world. Since Darwin, we've been a lucky chance. But until genomics we were still a beautiful oddity, a special case in nature. It's no trivial business that the twenty-first century is being shaped by the creation of artificial life, the engineering of existing life, and our dreams of escaping life altogether. These are disorienting times.

We now understand that the condition of a species is an almost magical weaving of time, a contemporary reality and a saga of metamorphosis. In this vision there's no evidence of the hard border between us and other animals of which we dream. Genes offer only change, mutation, disease and entanglement. Our physical form is porous, taking in scents, parasites, and even assimilating the DNA of other organisms. There's very little about us that suggests persistence. Our bodies are sublime, rebellious colonies of cells and our minds are floating, chameleon-like processes. This doesn't mean that we should see human life as meaningless. Thinking we are exceptional is different from thinking our lives have meaning. There's every reason to believe that our sense of significance is something we can't do without. But the weighing of human significance is less a fact of the world than a facet of our psychology. Where the difficulty lies is not in recognising ourselves as distinct creatures but in the ways our psychology builds its distinction. The trouble comes of trying to rubber-stamp our lives with proofs of exception.

An ancient question concerns the idea of pleasure. Why is pleasure so important to humans? Because pleasure is the avoidance of pain, Plato tells us. Philosopher Jeremy Bentham's rejoinder, centuries later, that the capacity to suffer can identify those to whom we have duties is only a reworking of old materials. It's clinically important and also captivating to observe the stunning complexity as pain lights up the amygdala in our brains, setting off bodily responses as part of a primitive means of defending the survival of organisms, and that related but in some ways distinct emotional narratives of suffering or danger swim out of our neocortices, blurred by old memories and knowledge, and become the fears we can express to one another. Current neuroscience tells us that only humans have this emotional story of fear. But we can't ever remove the fact that suffering and pleasure ultimately matter to our bodies. Even a more nuanced picture still boils down to the avoidance of pain or threat.

Language, too, has been a fickle foundation. It's fascinating to know that the Forkhead box protein P2, nicknamed FOXP2, encoded by the associated gene, is critical to normal development of speech and language in humans, and that we share this gene with a startling array of animals, including the robin singing in our back garden. Fascinating, too, that the gene is more active in females, and that only one amino-acid substitution differs between humans and chimpanzees. These are exhilarating discoveries, but they can't tell us much about why the possession of language should give us special rights.

What of Shamdeo, the wild boy found living with wolves on the far side of Musafirkhana forest in India, whom the inveterate wanderer Bruce Chatwin went to visit in 1978? 'He was always wary when entering a room, keeping to the wall and peering to see what lurked in the shadows,' wrote Chatwin. Shamdeo, like others of similar circumstances, never learned to speak, missing out on the first key stages of the window to acquire language that seems to close after around the first eight years of life. In what sense do we wish to think of Shamdeo as less human or important than us? The point is not whether a difference exists, as good as it is to know, but how we should think about a difference.

Many of the traits thought of as exclusively human were almost certainly present in other walking apes that are now extinct. The idea that a genetic thunderbolt gave us modern humans around 200,000 years ago is only another myth. Even the seemingly liberating idea that a cultural revolution took place a few tens of thousands of years ago is a story made from the faint echoes of a reality that may have been significantly different. The truth is we don't know how to separate this long and staggeringly old story such that some parts of the story somehow matter more than others.

The solution to this uncertainty has been to assume it's our moral faculty and our sparkling intellectual properties that raise the bar. But this is bias, not logic. The forefathers of modern science were convinced that rationality would usher in a sane and prosperous era. After all, only we can think of virtues or discuss ethics. As the only creatures that can do this, only we gain the special benefits that come along with it. Only those who can think of meaning can mean something. What they didn't bank on was that, for humans, being irrational is a sensible choice.

Whether soul or rationality or consciousness, these are spiritual or quasi-spiritual ideas that give a foundation for being good to one another. The basis of the protection and care of human beings still relies on a worldview that reinforces this assumption. Aquinas may have called it intrinsic worth. Kant may have repackaged it as human dignity, and so it goes on. Whatever it's called, the perspective is powerfully convincing. But that doesn't make it true. Insights into our behaviour and psychology present us with unique obligations towards one another. But none of this saves us from being animal. These days, humanists say that we should be defined as having minds filled with reasons. But they ignore the fact that we have reasons far more emotional than rational to believe this gives us impunity. Humans, contrary to the hopes of many, aren't shaped by moral logic but by an internal competition of possible choices. Our moral lives are an ad hoc mix of bias, intuition, conflicting histories of thought, and biological imperative.

In recent centuries, our societies appear to have gone through a moral awakening. Many countries have laws in place to protect the elderly and the vulnerable among us. While racism is still rife, more people are alert to it. Women, in a growing proportion of the world, are no longer under the absolute control of a patriarchal system. This is viewed as moral progress. Yet the awkward juxtaposition of the disappearance of diverse animal and plant life beside a heady optimism that this is the best era in our history should, at the least, awaken our curiosity. What might come in the aftermath of our celebrations as we appraise what we have done to life on Earth? Of course, there's nothing to stop us from dominating the Earth on our own terms and to hell with the rest of the living forms. Who knows, perhaps this is what any predator with the means would do. But we must give up the idea that we're moral.

Dreams of mutants and Mars

Imagine this future. It's 2320 and three cities have been built on Mars. The people living in this world have taken the genetic material they most prize from their origin planet and have fused it with synthetic DNA. There's a small forest through which the locals like to walk. There are no biting insects to trouble them. The bears that lumber about in the distance are docile, uninterested. The bees that wing across the bright flowers are stingless. There are no children bickering with each other in a flurry of rivalrous humour. The bees don't need to dance. The humans – if that's what we can call them – don't need to touch. Nothing has any intention of its own. There is no balance of strategies or intelligences. There isn't any need for other life forms to be intelligent. Evolution is a dirty word.

When we consider rebooting life on another planet, we become the creator of a new world. In setting ourselves up as creators, we begin to see the challenges of a new genesis. Why would we be taking life somewhere else? The stock answers to this assume that life generally has a value and that human life has a special value. But the arguments get much blurrier when it comes to the basis of the value. If life per se is a good thing, then in extending the biosphere by taking life elsewhere are we adding value to the universe? If it is life that has value, would Mars – a planet that we would people with some life forms, but not those we dislike – have slightly less value overall than an Earth with all its distinct and sometimes unpleasant inter-acting elements? With less overall diversity, would evolution on Mars be a lesser version of evolution on Earth? We like to think we're immune to this uncertainty, but any confusion surrounding the value of life more generally is a confusion in what grounds our own sense of meaning.

Colorado University astrobiologist Brian Hynek redirects our attention to the fact that we haven't yet decided 'if microbes matter'. Our psychology is biased against microbes because we see them as the source of threats like disease. On Earth, microbes are often seen as a pest, although that view is starting to shift as we learn more about their essential role in life. For Hynek, alien microbes are as thrilling as a whale swimming beneath the icy cover of Europa. Finding a single-cell organism on another planet would answer the question that we're not the only life in the universe, perhaps the most significant discovery in science. But in our fictional encounters with aliens, something always goes wrong. Most often our storytellers put us into contact with shadow selves. In our visions of alien life, we see what it's like to face a predator who only cares about itself.

In reality, Mars isn't a little Earth. Nor is anywhere else in our solar system. It's only our planet that has the perfect conditions for life. So, consider a matter closer to home. In 2016 the Australian Institute of Landscape Architects designed a beautiful, greener City of Melbourne, with living walls and breathing towers alive with flora. But if we invite nature back into our world, we will doubtless face pests both large and small. Do we modify our own skin to resist insect bites, or do we relax about the nuisance? Do we use gene-drives to remove species we don't like, or do we seek to mitigate but not to eliminate threats? Do we recognise the risk of predation of our family

pets, or do we pen them in? What relationship do we want with the rest of life?

For thousands of years we've tried to remake the Earth as we want it. But the fact is we're now killing it. Children around the world regard the biodiversity loss and destruction that we've caused as one of the most important issues in their lives. It has become normal to desire an ethic for our relationship to other species and their habitats. It has become normal to care. Yet the blind outcomes of evolution continue to unsettle us and worry at our efforts to make sense of the value of life. We attempt to shake this by trying to design life on our own terms.

But perhaps it is a mercy rather than a fault that there aren't intentions at work in the mechanisms by which life evolves. Last summer my sons and I spent a frenzied few weeks trying to see as many orchids as we could in the fields and woods near to our home. The species that took my children's interest was the bee orchid. After several efforts, we found one in an abandoned quarry twenty miles from our farm. When we arrived, there was nobody else around. Only marbled white butterflies trying to mate with each other, whirling above the grass tips like ashes from an unseen fire. The wasteland was a wen of orchids. Mostly common spotted and pyramidal. But after some careful hunting, we found a single bee orchid, tall and savvy.

What part of the evolving plant dreams this false bee into existence, having never seen one? What are we to make of an unthinking process that creates the image of a bee without the eyes to see it? These are wonders. Evolution is a wonder. But nothing evolves if there's only one set of motives in the rules of the game. If we design evolution by writing in our own intentions, we may discover the true horror of becoming a god.

The generous ape

The idea of generosity has swum through the books and notions of many thinkers over the centuries. It's there in most of the holy books. It crops up again and again in the works of philosophers. Schopenhauer suggested that the basis of morality was in the concepts of compassion and loving kindness, which he called *Menschenliebe*. Schopenhauer knew that generosity is closely associated with the ideas of love and feeling. The English word 'emotion' is related to the Latin *emovere*, which means 'to move away from'. In other words, these are good feelings that cause action. It is to this old meaning that we refer when we say that we are 'led by the heart'.

The word generosity itself has the whiff of superiority, emanating from *generosus*, or magnanimity, and referring to the higher nature of the nobility. Generosity, in this way, was the possession of superior emotions in a social hierarchy. But we can dig earlier to reclaim the word. It comes from the Old Latin *gegnere*, to beget or produce. Something is born that brings the gift of birth. To be generous in this way is to use a generous thought for an act of generosity. It's a mind that makes a mind in another. It's a feeling that sees into the feelings of something else.

In the ancient Daoist text of Zhuangzi, there is the well-known story of the Marquis of Lu, who attempts to revive a lost seabird by treating the creature as he himself would wish to be treated. The poor bird is given wine, the meat of a slaughtered bullock. In three days, it is dead. 'Had he treated it as the bird would like to be treated,' the author of the passage tells us, 'he would have put it to roost in a deep forest and allowed it to wander over the plain, swim in a river or lake, feed upon fish, and fly in formation with others.'

What the story shows us is that doing good is to treat something according to its own needs. It means that when we make choices about how to behave in relation to someone else, we can see into their needs and decide whether what we're doing will make us more generous towards those needs. The advantage here is that we respond to each individual as we find them, each species in its own measures, and each stage of a life cycle according to its nuance. The needs of a nuthatch are different from those of a human, just as a child's differ from those of an adult. Generosity is a flexible affair. The catch is that for this to be effective, it needs to tip more towards friendship than exploitation. A friendly relationship seems to be one in which something else exists not for us but for its own sake. That's the gold standard, whether it's a fellow human, a future child, or another living thing. There are plenty of mechanisms ready to counter this and disrupt it for the benefit of greater flexibility and individual gain. The one comfort comes in the fact that most of us pay a price for thwarting our generosity.

A modern worldview that divides us from the rest of life is a fundamentally ungenerous one. To support it, we have stripped the rest of the living world of its intelligence. When I was working in Boulder, Colorado, I shared a small house near a little dusty nature reserve. Opposite the reserve was an elementary school with a large playground in its field. While the schoolkids were on summer holidays, I used to take my children to the playground early in the morning to stop them waking anyone else in the household. The school field and the nature reserve were separated by a road that slid downhill in the direction of the city and the distant red smoke of the Rocky Mountains. While the kids yelled to their hearts' content, I would sit and stare across the road at the prairie dogs that lived in the reserve.

On one occasion, after heading back to the house, I came across the corpse of a young prairie dog that had been knocked down by an early-morning commuter. Somewhere out of sight, I could hear calls. Before long, another prairie dog arrived and moved around the corpse, crying out. It stayed there calling and gesturing in a way that was expressive enough to rattle me. Only a few days later, I was meeting with Marc Bekoff, professor emeritus of ecology and evolutionary biology at Colorado University. When I questioned him about prairie-dog behaviour, he directed me to something he'd written a year before. He had observed a similar incident with a black-tailed prairie dog, in which an adult attempted on five occasions to retrieve the body of a young prairie dog that had been run over, all the while 'emitting a very high-pitched vocalisation'.

Constantine Slobodchikoff began studying prairie dogs in the 1980s, when little was known about their social behaviour. It was understood that the animals split their 'towns' into territories with varying numbers. But these were presumed to be kin groups. When researchers began studying populations, they discovered instead that the size of these groups depended on the distribution of food, particularly whether it was evenly or patchily spread out, needing more or less defence. They also found that the groups included unrelated individuals who co-operated together. It had been assumed that the alarm calls made by the animals were all the same, a simple warning along the lines of 'watch out'. But when Slobodchikoff and his team began recording and analysing the calls, they found separate calls for a coyote, a human, a dog, a hawk, and so on. Not only that, but the acoustic structures of the calls resembled the phoneme units that make words, with information about colour, size and shape transmitted in the call.

Prairie dogs are a type of ground squirrel. For centuries, they've been killed because they are seen as competition on agricultural land. Today, they survive on about 2 per cent of their original range. Knowing what we do of the human mind, it shouldn't surprise us that we refuse to see into the needs of an obviously intelligent form of life when it gets in our way. But nor should it be a revelation that we can find it harder to kill them when we've looked into their lives more closely.

What human minds do so powerfully is exhibit a huge diversity of behaviours towards the same kinds of things, even to the same individuals. When the first humans – or the first primates – became hyperconscious persons, they uncovered ways to get along with each other. But humans have a social psychology that quivers between being defensive and supportive, collaborative and self-protective. The rub is that the very same spark that drives us into each other's arms and minds can bring out the worst in us. Our social minds draw us together as if we are in one mind on the world, but also protect us from those we don't wish to believe have much mind worth considering.

Believing there is something thinking and feeling inside an animal is an ancient means of causing a primate to be both cannier and more compassionate. We are people who give one another a piece of our minds to gain peace of mind. That all this may have been the consequence of a group of primates that needed one another to survive in no way belittles what it has left us with. Our intelligence made a discovery that doesn't belong only to us. Seeing into the minds or feelings of others and being generous made a new category of behaviour. We discovered how to think of others. We discovered that life, for all its hurt and predation, bears in its onward journey the kernel of generosity. We can even think into what the wants, if we wish to call them this, of an ecosystem might be. When our minds generated insights, love escaped the bounds of self, of kin, of neighbour, and spread outwards.

What seems clear is that humans need meaning to reassure us in a world in which we have to die. Modern societies have been built on origin stories that tell us we are the highest life form in an evolutionary hierarchy. But it might be preferable to overcome the use of superiority to structure justice. One way to be reconciled with a reality that disturbs us is to reconsider it in a more convivial light. It's possible that when we see the world as alive with intelligence, it doesn't seem such a threatening place. Albert Einstein was onto this more than half a century ago. In a letter written to an individual who had suffered a death in the family, he wrote: 'A human being is a part of the whole called by us universe, a part limited in time and space. He experiences himself, his thoughts and feelings as something separated from the rest, an optical illusion of his consciousness.' Our task is to free ourselves from these limits, as 'only overcoming it gives us an attainable measure of inner peace'.

Our value systems place constraints on our behaviour. To help us to stop over-utilising our planet in a way that is detrimental to life as a whole, we may need to light on the feelings, sensations, intelligence and intentions of the other lives around us. And, as we see into these lives, there's nothing to stop us extending intrinsic value to them. Perhaps intrinsic value is only a human concept. Or perhaps it's the way an animal comes to care for another animal without conditions. Arbitrary? Well, perhaps. But why not? We ourselves don't really fit into the the neat thresholds of value that we have created up to now. To safeguard one another, we have come up with the great conceptual leveller of 'dignity'. Dignity encourages us – and with good reason – to see each human life, regardless of any biological differences in stage of life or capacities – as precious and worthy of respect. Dignity, in many ways, is a powerful form of insight.

But there's nothing that should hold us back from seeing into other species. All forms of life are exceptional. Life, for goodness' sake, is exceptional. Whatever specific obligations we have towards one another, they don't divide us from the rest of life. Nor do they guarantee we are alone in our worth. Our attributes undoubtedly give us an amazing range of behaviours, but this versatility and resilience doesn't bring our animal life to an end. Our proper place is with our fellow creatures. It's time we told ourselves a new story of revolutionary simplicity: if we matter, so does everything else.

CODA: ON THE LOVELINESS
OF BEING ANIMAL

. . . to put a hand on its brow
of the flower
and retell it in words and in touch
it is lovely
until it flowers again from within, of self-blessing.

Galway Kinnell

The other night I experienced a feeling that came out of nowhere. Of course, that's the wrong turn of phrase. It came out of nothing I'd been aware of thinking about. Suddenly I began to cry. Out loud. This is unlike me. This feeling swept out of my body like the landfall of a wind that has gathered its energy out at sea. In an instant, I knew quite lucidly that this was the appearance of grief at the knowledge that my child, nearly ten now, is on the threshold of independence. He is climbing, without even knowing it, out of the gardens of our shared experience of his childhood. Slowly he will visit this shared space less and less often as he journeys into whatever life has in store for him. It was a brief singularity of loss – of time, of lost moments together, of my ability to protect him.

Inside the feeling, perhaps too, a little sorrow at the hurrying along of my own life and my own lost childhood.

I emailed my mother the next day. Her response was immediate. 'I'm afraid the waves of sadness will go on happening,' she wrote, 'and you will get them as a grandmother too! You wish you could make it last just a little bit longer. But it gets snatched away from us, and time propels us relentlessly along.' Imagine, though, if I could live another fifty years. Imagine if I could have several children with several different partners in separate stages through my extended life, grafting the virginal slices of my younger self's ovaries to delay menopause, or perhaps growing a whole new organ of reproduction. Always delaying loss and acquiring just a little more for myself. Or imagine that we let go of the promise of medicine and turn more and more of ourselves over to machines. We stand on the brink of choices like this. But we also see before us a multiplex of other animals, plants and organisms whose lives are in thrall to ours. The more we take, the more their lights go out. What will we make of this world of ours?

Suspending disbelief for just a moment, what would we gain if we weren't an animal? Or, if we flip the question, what do we gain from being an animal? Think of music. In surveys, around 90 per cent of us listen to music for many hours of every week of our lives. A life without music would be an impoverished one for many. For a long time, researchers hoped to find a 'music bit' of the brain to explain what seemed a peculiarly human behaviour. No such luck. In a recent study across Chinese and American participants, at least thirteen overarching emotions were activated by music, many of which rely on different aspects of our biology. The great waves of sadness and joy, exhilaration and pleasure we experience when we listen to music pull from multiple aspects of being an animal.

Not only that, but they don't just unfold in the brain or rely on our sense of self. They're modulated by organic chemicals and processes that act like filigrees of feeling, tied throughout the body in a net of impossible complexity. To be frank, music depends on us being animal.

So how have we come to think of music as a matter of mathematics and algorithms? A computer might be able to retrieve the building blocks of musical systems from the work of millions of composers and knock out a passable replica of the work they've done. But music only matters if we can feel it. Listening to music has more to do with the scent of honey or the pleasure of reproducing or of being healthy in a plentiful landscape than it does with a mathematical algorithm. For that matter, responding to music, whether it's Mozart or Lady Gaga, has more to do with the experiences of whales deep in the Atlantic Ocean than it does with the computer through which we might be listening.

We might argue that it is our identity and memories that make music so special to us. But being a person is really a memory of being an animal. In English, the word 'meaning' derives from an older word *meninge*, the word that is now used to describe parts of our brains (and diseases that destroy our brains, such as meningitis). The earlier sense of this word in Latin was a kind of remembering. Meaning, then, is an act of recall. That becomes more than ironic when human culture makes the decision to forget that we are animals.

The memories we retrieve are the experiences of stimuli at the cellular level throughout our bodies. As the world is an onrush of sensations that every part of our flesh mops up – in scent, sound, taste, touch and smell – memories are much more than our thoughts about things. They are the pleasures and anguish, the pains and delights that we undergo as animals. And so, as we search for the meaning of our lives, we might ask ourselves whether being animal is the meaningful thing after all.

Before we rush to rid ourselves of the burden of creature-liness, think on this. Some of the most important phases of our lives take place in the rhythmic dusk of our mother's womb and in her arms in the first stages after birth. The bonds that bring us together in friendships, romantic relationships of any sexual preference, and in our tolerance of those we barely know, are partly built on physical systems in our bodies that are shaped by the ancient alliances between parents and their offspring. Researchers such as Israeli psychologist Ruth Feldman have revolutionised our insights into early development. 'Later attach-ments in all their forms repurpose the basic machinery established by the mother-offspring bond during sensitive periods,' she writes.

The work to unravel some of this began in studies on rats and the monogamous prairie voles. The life of a young mammal opens with a helpless infant that needs to be close to a nursing mother. We now know that there's crosstalk of oxytocin and dopamine, and that pivotal phases in the growth and nurturing of a foetus and infant alter how the brain's neural networks are or-ganised. Across vertebrate

evolution, the molecule from which oxytocin comes has fiddled with fundamental life controls to give us a wide variety of possible social behaviours. Life is nothing if not thrifty. In mammals, dopamine in the nucleus accumbens region of the brain is linked to oxytocin receptors. The greater the association between oxytocin and dopamine, the more the reward systems in the brain can be flexible and allow for the making of new alliances. In one study, the density of oxytocin receptors in developing prairie voles was linked to the amount of time the individual later spent huddling with a partner.

But the bond between mothers and infants among rodents is brief. It lasts as long as it's needed. Humans are different again. Primates like us live in groups with long-term, changeable relationships, defined by memories and hormones and observations of one another. Nevertheless, the mother's body remains a crucial environment for social development. Inside her, a growing child comes to know the rhythm of her heart and the amount of stress her body endures. Outside her, a baby learns how she touches, how she responds to others socially, and how stressed she is. This in turn alters the development of the child's central nervous system and modulates the brain.

Through the early years of our lives, our bodies form sensual and physical memories of our experiences in attachments that take us through all the other phases of the human life cycle. And for humans, the presence and parenting behaviour of the father or partner and other close individuals also play a signifi-cant role in infant brain development as well as in reducing the stress of the primary caregiver. While we have the gift to repair early bad relationships, those first positive affiliations and alliances can exert an influence on our capacity for compassion throughout our lives. A recent suggestive finding has shown how oxytocin affects the amygdala in mammals. When mammals have good early attachments, these expose the amygdala to long-term oxytocin, lowering the fear response to dangerous or unpleasant situations and improving the extent to which positive attachments can be used to benefit from relationships in the future.

In all this, humans are remarkable. From these early relationships we go on to make others of many and various kinds. It isn't unreasonable to suggest that our ability to bond with everything from a stranger on a train journey to a chinchilla is the most exciting aspect of our psychology. The flexibility of our attachments and the possible relationships we can form are unusually wide. Our love and consideration of one another can travel through time and into symbolic worlds. We can love the dead and the idea of a country. We can love God and the bird in the back garden. And yet each of these different kinds of love make use of the ways our bodies form bonds in infancy and are influenced by the early attachments we witness and experience as a child. We are only at the beginnings of discovering how flexible this model of generating life might be. How much of our sexual history can we engineer to suit the purposes of living adults? Right now, we have no idea.

It is only when we imagine cutting away the body or simulating it that we glimpse the extravaganza of what it does. We begin as a bunch of splitting and proliferating cells in the darkness of a womb. Soon we will become an aquatic creature, bobbing about in a private ocean. In only a few weeks a primitive face will form, a tiny, hairbreadth mouth will grow. In a month or so, little hands the width of an eyelash will begin to clench and unfold. While in the womb, some of our cells will migrate into our mother's body. The process is called microchimerism and it's widespread among mammals. These cellular ghosts of children remain in the mother's body for the rest of her life.

Out of the womb, babies seek out their mother's nipple. Flesh of child and flesh of mother touch. Fatty acids in the milk, along with antibodies and bactericidal protein, spray into the throat of the baby, transferring nutrition and protection against disease. At the same time, the mother and the child experience surges in oxytocin, with profound effects on mood, social behaviour, sense of smell and responses to stress. This affects the amygdala in the brain, which then adjusts our attention and responses to our children. We stare at our children's faces and the reward centres of our brains glint in the dark.

As newborns, we answer even at the molecular profile of our cells to the amount our mothers touch us. The embrace of the parent and their child regulates the baby's body temperature, heart rate and respiration. When we hold a child that is sad or frightened, we slowly stroke their backs without even thinking about it. Their pulse slows, the sobs slow, the breathing slows, and slowly they return the touch, until the sun comes out again. This is life in the flesh.

There's extraordinary behavioural synchrony that we find among humans, something absent from an interaction with a computer or a phone. People in an alliance, however temporary, mirror one another's touch and body language. Lovers can synch the gamma oscillations and certain hormones in their bodies when they're together. But it's most powerful between a mother

and her child in the early stages of life. When their bodies are in harmony with each other, their heart rates can synchronise, the oscillations of alpha and gamma rhythms in their brains can come together, and oxytocin and cortisol levels begin to shift.

And it isn't just us. In other animals, their social behaviour and a range of bodily functions are shaped by the amount of touch they receive. What we have in common with all mammals are what is called C tactile afferents. These are neurons found in hairy skin and the spinal cord. The amount that mother rats lick their offspring has a radical effect on the physiology of the babies. It is safe to assume that similar mechanisms, along with the pleasant feeling they can unleash and the advantage they can confer on the body, are shared across the mammals on our planet.

In the 1950s, psychologist Harry Harlow took baby rhesus monkeys from their mothers at birth, isolating some of them for as much as a year. He wanted to prove the importance of tactile comfort for behaviour. He found that once returned to the company of other monkeys, the individuals were unable to socialise, tore their fur out and bit their own arms and legs. Unsurprisingly, the effects were permanent among those that had been isolated the longest. Later he studied infant monkeys that were given two mother models, a wire mesh that provided milk and a soft cloth mother. What he found was that the monkeys only went to the wire mother for food, returning to cuddle the cloth dummy. After more than a decade of experiments, he came to the unremarkable conclusion that physical, social affection and interaction are critical to the normal development of the monkeys, especially in the first months. In so doing, he succeeded in telling us what mothers have known for thousands of years.

Under sane conditions, such studies are not permitted on people. The only similar experiment sanctioned in human models in recent history took place when former Romanian leader Nicolae Ceauçescu saw fit to abandon one hundred thousand children in orphanages. For Ceauçescu, the horrific disregard for the lives of the neglected children was made possible by the ruthless demands of his grandiosity. Raised without sensory stimulation or physical affection, the children showed a range of severe physical struggles, from balance and coordination to visual-spatial understanding and language delays.

The painful truth at the core of such experiments is that our bodies may be fundamental to who we are. This isn't to say that there's something essentially wrong with altering our bodies in our lifetimes. Saying our bodies may be fundamental is only to suggest it may not be possible to transition from being a human animal on Earth in any meaningful way. Doubtless we can modify and extend the capacities of our physical forms and our lifespans. But there may come critical points beyond which the dynamism of life won't bend to our will. The personal view we experience is significantly affected by everything from how much sleep we've had, to the rise and fall of hormones, to a bleed on the brain. To argue otherwise is to dream that we are as rigid as a computer program. It is to wish for us to be fixed somehow.

A specialist in artificial intelligence, Peter Bentley, thinks we should be wary of assuming machine intelligence can replace rather than supplement human and animal intelligence. 'We assume that intelligence arises from complexity, but the model might be wrong,' he told me. It may, for instance, emerge through something like homeostasis. 'Evolution has had four billion years' worth of testing,' Bentley says. 'Evolution itself manipulates genes as a kind of intelligence, solving the problems of survival.' Do we really want to deskill evolution? The danger is that we will 'remove the thought process' from life.

After all, a computer is supposed to be a tool, not a person. It should assist, not replace us. If we use a computer or a phone to do the work of a parent or friend in the early stages of a child's development, we assume that intelligence lives only in the brain. We forget that when people learn from one another, they not only exchange information that will improve their knowledge, they also exploit every aspect of their bodies to extend learning into movement, social intelligence, long-term memory and self-understanding.

What happens if we replace the flesh that came into being on its own terms with metals or materials that we mine from an asteroid owned by a private company? Computer programs can be hacked, deleted and rewritten. Compared to synthetic models of intelligence, our bodies are much more invincible. Our conscious experience can lessen or disappear under certain limitations, yet the canny body remains. Even when Alzheimer's releases its acid through the memories and quirks of the self, for a while the body keeps on its intelligence, the gut does its work, the blood and cells and enzymes keep busy. From this perspective, we seem to be the temporary watchers of a life force that somehow knows what to do in our absence. It is as though there's a more authoritative wit throughout the living systems of our bodies to which we are fleetingly privy.

We're hellbent on thinking about what machines can do at the cost of noticing how awesome is the human body that made them. All of us panic from time to time about illnesses. Many of us rely on our health services to stay alive and well. But we forget that doctors are often only supporting our incredible natural abilities. Our immune system is made up of so many discrete elements that we can be beguiled into forgetting it's there. Yet what a thing it is. The cells and chemistry inside our bodies must distinguish which stuff is us from which isn't, and what is harmless from what could kill. It took scientists decades to establish how animals undergo immune responses to novel encounters with manmade chemicals that wouldn't have been in the environment of adaptation. How did the aspects of the immune system come to know something unknown, let alone classify it as a danger? What immunologists discovered was that there are capacities to probe and identify almost on a molecular scale. And our bodies can do this because all organisms on our planet are fundamentally similar. It's kinship that makes our immune system work.

'I believe in you my soul,' wrote Walt Whitman in his 'Song of Myself', 'the other I am must not abase itself to you./And you must not be abased to the other.' In Whitman's vision, it makes no sense for the thinking self to transcend its animal body. Without the body, the soul is a pointless abstraction. When we try to save the person by ceasing to be animal, we forget that a person and an animal are the same thing. There's nothing to escape. We are already what we should be. And, in most ways that matter, we are already what we want to be. When we look up at the stars on a crisp dark night, we not only see the alchemy of light turning into memory, but our body also remembers a time without light. As our eyes adjust to the dark, they switch from cone to rod cells, an adaptation that can take nearly an hour. Photons of light interact with protein molecules in the photoreceptors of our eyes. We share the same essential molecule in our photoreceptors as all other vertebrates on Earth. Our eyes gaze on through the haze of kindredship.

For now, humans are Earth creatures, still born of flesh, conscious, emotional and mortal. We have laptops, smart-phones, here or there a titanium heart valve, a prosthetic limb or exoskeleton where machine and nerves fuse. We have high-rises, rockets and space stations. But, still, we remain animals. We are the selfsame star-stuff we have always been. As animals, we harbour viruses. We fart out the digestive gases from our meals. Our plump, bloody hearts drum every feeling we have. As we grow, we seek a mate. Staring into their eyes, our bodies increase the blood flow into the nucleus accumbens region of our brains, the same region that comes to life when mothers look at their babies. When we sniff the heads of our newborn children, we change the hormones in our bodies and love breaks out like a rash. When we stroke their bodies, we grow kindness in them. When a friend is in need, we make them laugh and endorphins rush through us as hope. The subjective consciousness that slowly develops through childhood and young adulthood and passes away again in old age is only a stage in the unfolding of a human life.

If our bodies count, if they possess their own intelligence, if they are the theatre through which our sense of self occasionally flickers like a faulty connection in the wiring, then dramas of beauty and meaning may continue even in the darkness. This is a radical possibility. If much of the real art is in the body and not the knowledge of the body, then while moral agency may be limited, moral subjects may be everywhere. We can imagine not a progressive scale leading to the one thing of value – the human form of consciousness and its capacity to deliver meaning – but a spectacle of richness before us all the while.

SELECTED BIBLIOGRAPHY

My thinking for this book stretches back well over a decade. As such, there are a wide number of sources I've consulted and researchers whose work has been significant to me. I also quote from a lot of recent studies. I've tried to ensure that the science enhances the central ideas but that the main claims – that humans are animals, that we struggle with that fact, and that this matters to us profoundly – can't be undermined by any science that might turn out to need a slightly different interpretation to that we now give. Below is a list of the papers or articles from which I have quoted, for those who would like to trace back to the original source and read the work in full. I've also, where useful, included relevant books by the researcher or thinker.

Adler, Mortimer J., *The Difference of Man and the Difference It Makes* (Fordham University Press, 1993)

Ariely, Dan, *The Honest Truth About Dishonesty: How We Lie to Everyone – Especially Ourselves* (HarperCollins, 2012)

Bain, Paul G., Jeroen Vaes and Jacques-Philippe Leyens (eds), *Humanness and Dehumanization* (Psychology Press, 2013)

Bandura, Albert, *Moral Disengagement: How People Do Harm and Live With Themselves* (Worth Publishers, 2016); also Bandura, Albert, Claudio Barbaranelli, Gian Vittorio Caprara and Concetta Pastorelli,

'Mechanisms of Moral Disengagement in the Exercise of Moral Agency' (*Journal of Personality and Social Psychology*, 1996)

Banks, Iain M., *Surface Detail* (Orbit, 2010)

Barash, David, 'It's Time to Make Human–Chimp Hybrids: The Humanzee Is Both Scientifically Possible and Morally Defensible' (March, 2018). It can be found here: nautil.us/issue/58/self/its-time-to-make-human_chimp-hybrids; also his book *Through a Glass Brightly: Using Science to See Our Species As We Really Are* (Oxford University Press, 2018)

Bar-On, Yinon M., Rob Phillips and Ron Milo, 'The Biomass Distribution on Earth' (*PNAS*, 2018)

Barron, Andrew and Colin Klein, 'What Insects Can Tell Us About the Origins of Consciousness' (*PNAS*, 2016)

Beatson, Ruth, Michael Halloran and Stephen Loughnan, 'Attitudes Toward Animals: The Effect of Priming Thoughts of Human-Animal Similarities and Mortality Salience on the Evaluation of Companion Animals' (*Society and Animals*, 2009)

Bednarik, Robert G., 'The Origins of Symboling' (*Signs Journal*, 2008); also 'On the Neuroscience of Rock Art Interpretation' (*Time and Mind*, 2013)

Bickerton, Derek, *Adam's Tongue: How Humans Made Language, How Language Made Humans* (Hill & Wang, 2009)

Braidotti, Rosi, *The Posthuman* (Polity Press, 2013)

For the new technique of transparency, see Cai, Ruiyao et al., 'Panoptic Imaging of Transparent Mice Reveals Whole-Body Neuronal Connections' (*Nature Neuroscience*, 2018)

Chalmers, David and Andy Clark, 'The Extended Mind' (*Analysis*, 1998)

Church, George and Ed Regis, *Regenesis* (Basic Books, 2012)

Clayton, Nicola S. and Anthony Dickinson, 'Episodic-like Memory During Cache Recovery by Scrub Jays' (*Nature*, 1998); also Emery, Nathan J. and Nicola S. Clayton, 'The Mentality of Crows: Convergent Evolution of Intelligence in Corvids and Apes' (*Science*, 2004)

For the paper on IVG, see Cohen, Glenn, George Q. Daley and Eli Y.

Adashi, 'Disruptive Reproductive Technologies' (*Science Translational Medicine*, 2017)

Corballis, Michael C., 'Wandering Tales: Evolutionary Origins of Mental Time Travel and Language' (*Frontiers in Psychology*, 2013); also Suddendorf, Thomas and Michael Corballis, 'Mental Time Travel and the Evolution of the Human Mind' (*Psychological Monographs*, 1997) and Corballis, Michael C. and Thomas Suddendorf, 'Memory, Time, and Language' in C. Pasternak (ed.), *What Makes Us Human* (Oneworld Publications, 2007)

Costello, Kimberly and Gordon Hodson, 'Exploring the Roots of Dehumanization: the Role of Animal-Human Similarity in Promoting Immigrant Humanization' (*Group Processes and Intergroup Relations*, 2010)

For the study on Cova dels Cavalls, see López-Montalvo, Esther, 'Hunting Scenes in Spanish Levantine Rock Art: An Unequivocal Chrono-cultural Marker of Epipalaeolithic and Mesolithic Iberian Societies?' (*Quaternary International*, 2018)

Cremer, Sylvia, Sophie A. O. Armitage and Paul Schmid-Hempel, 'Social Immunity' (*Current Biology*, 2007)

Cummins, Denise, 'Dominance, Status, and Social Hierarchies' in Buss, David M. (ed.), *The Handbook of Evolutionary Psychology* (Wiley, 2005)

Damasio, Antonio, *Self Comes to Mind* (Pantheon, 2010)

Davis, Matt, Søren Faurby and Jens-Christian Svenning, 'Mammal Diversity Will Take Millions of Years to Recover from the Current Biodiversity Crisis' (*PNAS*, 2018)

De Dreu, Carsten K. W. and Mariska E. Kret, 'Oxytocin Conditions Intergroup Relations Through Upregulated Ingroup Empathy, Cooperation, Conformity, and Defense' (*Biological Psychiatry*, 2016)

De Pasquale, Concetta et al., 'Psychopathological Aspects of Kidney Transplantation: Efficacy of a Multidisciplinary Team' (*World J. Transplant*, 2014)

Della Porta, Donatella, 'Radicalization: A Relational Perspective' (*Annual Review of Political Science*, 2018)

Dorovskikh, Anna, 'Killing for a living: psychological and physiological effects of alienation of food production on slaughterhouse workers'. Undergraduate honours thesis, University of Colorado, Boulder.

Du, Hongfei, et al., 'Cultural Influences on Terror Management: Independent and Interdependent Self-esteem as Anxiety Buffers' (*Journal of Experimental Social Psychology*, 2013)

Eisnitz, Gail A., *Slaughterhouse: The Shocking Story of Greed, Neglect, and Inhumane Treatment Inside the U.S. Meat Industry* (Prometheus, 2009)

England, Jeremy, 'A New Physics Theory of Life' (*Quanta* magazine, 2014). To read more on Jeremy England's theories, visit www.quantamagazine. org/first-support-for-a-physics-theory-of-life-20170726/

Robert S. Feldman's work on lying can be found in Moskowitz, Michael, *Reading Minds: A Guide to the Cognitive Neuroscience Revolution* (Routledge, 2018)

Feldman, Ruth, 'Maternal and Paternal Plasma, Salivary and Urinary Oxytocin: Considering Stress and Affiliation Components of Human Bonding' (*Developmental Science*, 2010)

Fernandez, Silvia, Emanuele Castano and Indramani Singh, 'Managing Death in the Burning Grounds of Varanasi, India: A Terror Management Investigation' (*Journal of Cross-cultural Psychology*, 2010)

Fitzgerald, Amy J., Linda Kalof and Thomas Dietz, 'Slaughterhouses and Increased Crime Rates: An Empirical Analysis of the Spillover from "The Jungle" into the Surrounding Community' (*Organization and Environment*, 2009)

Sarah Franklin has written extensively on biotechnology, and her works are available on her website, sarahfranklin.com. To see the text I used, visit sarahfranklin.com/wp-content/files/Franklin-Life-entry-from-Bioethics-2014.pdf

Frith, Chris D. and Uta Frith, 'Interacting Minds – A Biological Basis (*Science*, 1999); also Frith, Uta and Chris D. Frith, 'Development and Neurophysiology of Mentalizing' (*Philosophical Transactions of the Royal Society of London, Series B*, 2003)

Gaesser, Brendan and Daniel L. Schacter, 'Episodic Simulation and

Episodic Memory can Increase Intentions to Help Others' (*PNAS*, 2014)

Gagneux, Pascal, Christophe Boesch and David S. Woodruff, 'Female Reproductive Strategies, Paternity, and Community Structure in Wild West African Chimpanzees' (*Animal Behaviour*, 1999)

Goldenberg, Jamie L. and Kasey Lynn Morris, 'Death and the Real Girl: The Impact of Mortality Salience on Men's Attraction to Women as Objects' in Roberts, Tomi-Ann et al. (eds), *Feminist Perspectives on Building a Better Psychological Science of Gender* (Springer, 2016); also Cox, C. R., J. L. Goldenberg, J. Arndt and T. Pyszczynski, 'Mother's Milk: An Existential Perspective on Negative Reactions to Breastfeeding' (*Personality and Social Psychology Bulletin*, 2007)

Haidt, Jonathan et al., 'Moral Foundations Theory: The Pragmatic Validity of Moral Pluralism' (*Science*, 2012)

Haile-Selassie, Y., S. M. Melillo, A. Vazzana, S. Benazzi and T. M. Ryan, 'A 3.8-million-year-old Hominin Cranium from Woranso-Mille, Ethiopia' (*Nature*, 2019)

Harris, Lasana T. and Susan T. Fiske, 'Dehumanizing the Lowest of the Low: Neuroimaging Responses to Extreme Out-Groups' (*Psychological Science*, 2006); also Harris, Lasana T., *Invisible Mind: Flexible Social Cognition and Dehumanization* (MIT Press, 2017)

To read more on He Jiankui, see Cohen, Jon, 'The Untold Story of the "Circle of Trust" Behind the First Gene-edited Babies' (Science, 2019) accessible at www.sciencemag.org/news/2019/08/untold-story -circle-trust-behind-world-s-first-gene-edited-babies

Heilbroner, Robert, *Visions of the Future: The Distant Past, Yesterday, Today, and Tomorrow* (Oxford University Press, 1996)

Henrich, Joseph, *The Secret of Our Success: How Culture Is Driving Human Evolution, Domesticating Our Species, and Making Us Smarter* (Princeton University Press, 2015)

Herzog, Harold A., 'The Moral Status of Mice' (*American Psychologist*, 1988); see also Herzog, Hal, *Some We Love, Some We Hate, Some We Eat: Why It's So Hard to Think Straight About Animals* (HarperCollins, 2010)

Hill, Kim, Michael Barton and A. Magdalena Hurtado, 'The Emergence of Human Uniqueness: Characters Underlying Behavioral Modernity' (*Evolutionary Anthropology*, 2009)

Hirschberger, Gilad, Victor Florian and Mario Mikulincer, 'The Existential Function of Close Relationships: Introducing Death into the Science of Love' (*Personality and Social Psychology Review*, 2003)

Hodgson, Derek, *The Roots of Visual Depiction in Art: Neuroarchaeology, Neuroscience and Evolution* (Cambridge Scholars, 2019)

Hrdy, Sarah Blaffer, *Mothers and Others: The Evolutionary Origins of Mutual Understanding* (Harvard University Press, 2011)

Impey, Chris, *Beyond: Our Future in Space* (W. W. Norton, 2015)

Jackendoff, Ray, 'Your theory of language evolution depends on your theory of language' in Larson, Richard K., Viviane Déprez and Hiroko Yamakido (eds), *The Evolution of Human Language: Biolinguistic Perspectives* (Cambridge University Press, 2010); also *Foundations of Language: Brain, Meaning, Grammar, Evolution* (Oxford University Press, 2002)

Jasanoff, Alan, *The Biological Mind: How Brain, Body, and Environment Collaborate to Make Us Who We Are* (Basic Books, 2018)

Jost, John T. and George A. Bonanno, 'Conservative Shift Among High-Exposure Survivors of the September 11th Terrorist Attacks' (*Basic and Applied Social Psychology*, 2006)

Keil, Frank, *Concepts, Kinds, and Cognitive Development* (MIT Press, 1989)

Kipnis, David, Patricia J. Castell, Mary Gergen and Donna Mauch, 'Metamorphic effects of power' (*Journal of Applied Psychology*, 1976)

Kurzweil, Ray, *How to Create a Mind: The Secret of Human Thought Revealed* (Duckworth, 2014); also kurzweilai.net

Lane, Nick, *The Vital Question: Why Is Life the Way It Is?* (Profile, 2015)

Levy, Jonathan et al., 'Adolescents Growing Up Amid Intractable Conflict Attenuate Brain Response to Pain of Outgroup' (*PNAS*, 2016)

Llinás, Rodolfo, U. Ribary, D. Contreras and D. Pedroarena, 'The Neuronal Basis of Consciousness' (*Philosophical Transactions of the Royal Society of London Series B*, 1998)

The Nick Longrich quote comes from his article 'Were Other Human Species the First Victims of the Sixth Mass Extinction' on realclearscience.com (November 2019) accessible at www.realclearscience.com/articles/2019/11/23/were_other_human_species_the_first_victims_of_the_sixth_mass_extinction_111191.html

MacIntyre, Alasdair, *Dependent Rational Animals* (Bloomsbury Academic, 2009)

Malotki, Ekkehart and Ellen Dissanayake, *Early Rock Art of the American West: The Geometric Enigma* (University of Washington Press, 2018)

McBrearty, Sally and Alison S. Brooks, 'The Revolution That Wasn't: A New Interpretation of the Origin of Modern Human Behavior' (*Journal of Human Evolution*, 2000)

Melis, Alicia P., Patricia Grocke, Josefine Kalbitz and Michael Tomasello, 'One for You, One for Me: Humans' Unique Turn-taking Skills' (*Psychological Science*, 2016)

Menzel, Randolf, G. Leboulle and D. Eisenhardt, 'Small Brains, Bright Minds' (*Cell*, 2006)

Meunier, Joël, 'Social Immunity and the Evolution of Group Living in Insects' (*Philosophical Transactions of the Royal Society of London, Series B*, 2015)

Midgley, Mary, 'Biotechnology and Monstrosity: Why We Should Pay Attention to the "Yuk Factor"' (Hastings Center Report, 2000); *Beast and Man: The Roots of Human Nature* (Cornell University Press, 1978)

Molenberghs, Pascal et al., 'The Influence of Group Membership and Individual Differences in Psychopathy and Perspective Taking on Neural Responses When Punishing and Rewarding Others' (*Human Brain Mapping*, 2014); also Molenberghs, Pascal and Winnifred R. Louis, 'Insights from fMRI Studies into Ingroup Bias' (*Frontiers in Psychology*, 2018)

Mormann, Florian et al., 'A Category-specific Response to Animals in the Right Human Amygdala' (*Nature Neuroscience*, 2011)

Motyl, Matt et al., 'Creatureliness Priming Reduces Aggression and Support for War' (*British Journal of Social Psychology*, 2012)

The source of the quote by Elon Musk on 'summoning a demon' is in his speech at the 2014 MIT Aeronautics and Astronautics Department's Centennial Symposium. Visit: www.youtube.com/watch?time_continue=5&v=zQcNdEMJ38k&feature=emb_title

Nagasawa, Miho et al., 'Oxytocin-gaze Positive Loop and the Coevolution of Human–Dog Bonds' (*Science*, 2015)

Navarrete, Carlos, 'Death Concerns and Other Adaptive Challenges: The Effects of Coalition-relevant Challenges on Worldview Defense in the US and Costa Rica' (*Group Processes and Intergroup Relations*, 2005); also Navarrete, C. D. and D. M. T. Fessler, 'Normative Bias and Adaptive Challenges: A Relational Approach to Coalitional Psychology and a Critique of Terror Management Theory' (*Evolutionary Psychology*, 2005)

Nussbaum, Martha C., 'Objectification' (*Philosophy & Public Affairs*, 1994); also *Upheavals of Thought: The Intelligence of Emotions* (Cambridge University Press, 2001)

Olson, Eric, 'An Argument for Animalism' in Martin, R. and J. Barresi (eds), *Personal Identity* (Blackwell, 2003); also Blatti, Stephen and Paul F. Snowdon (eds), *Animalism: New Essays on Persons, Animals, and Identity* (Oxford University Press, 2016)

Oppenheimer, Stephen, *The Real Eve: Modern Man's Journey Out of Africa* (Carroll & Graf, 2004)

Pepperberg, Irene M., 'Cognitive and Communicative Abilities of Grey Parrots' (*Current Directions in Psychological Science*, 2002); also *Alex & Me* (Collins, 2008)

For the section on autotrophs, I was inspired by a sobering reflection on energy by David Price, 'Energy and Human Evolution' (*Population and Environment*, 1995)

For the source of the quote from Giulio Prisco, see *Tales of the Turing Church: Hacking Religion, Enlightening Science, Awakening Technology* (independently published, 2018)

Pyszczynski, Thomas, Jeff Greenberg and Sheldon Solomon, 'A Dual Process Model of Defense Against Conscious and Unconscious

Death-related Thoughts: An Extension of Terror Management Theory' (*Psychological Review*, 1999); see also *The Worm at the Core: On the Role of Death in Life* (Penguin, 2016)

Rabasa, Angel et al., *Beyond Al-Qaeda* (RAND Corporation, 2006)

Rizal, Yan et al., 'Last Appearance of Homo erectus at Ngandong, Java, 117,000–108,000 Years Ago' (*Nature*, 2019)

Roosth, Sophia, *Synthetic: How Life Got Made* (University of Chicago Press, 2017)

Roylance, Christina, Andrew A. Abeyta and Clay Routledge, 'I Am Not an Animal But I Am a Sexist: Human Distinctiveness, Sexist Attitudes Towards Women, and Perceptions of Meaning in Life' (*Feminism and Psychology*, 2016)

Sandel, Michael J., *The Case Against Perfection* (Belknap Press, 2007)

Savulescu, Julian, 'In Defence of Procreative Beneficence' (*Journal of Medical Ethics*, 2007)

Schacter, Daniel L., Joan Y. Chiao and Jason P. Mitchell, 'The Seven Sins of Memory: Implications for the Self' (*Annals of the New York Academy of Sciences*, 2003)

Schrödinger, Erwin, *What Is Life?* Published in 1944, it can be downloaded online and Cambridge University Press published a new edition in 2012.

Schwab, Klaus, *The Fourth Industrial Revolution* (Portfolio/Penguin, 2017)

Smith, David Livingstone, *Less Than Human: Why We Demean, Enslave and Exterminate Others* (Griffin, 2012)

Thacker, Eugene, *Starry Speculative Corpse* (Zero Books, 2015)

Thomas, James and Simon Kirby, 'Self-domestication and the Evolution of Language' (*Biology and Philosophy*, 2018)

Tomasello, Michael and Henrike Moll, 'The Gap Is Social: Human Shared Intentionality and Culture' in Kappeler, Peter M. and Joan B. Silk (eds), *Mind the Gap: Tracing the Origins of Human Universals* (Springer, 2009)

Tooby, John and Leda Cosmides, 'Groups in Mind: The Coalitional Roots of War and Morality' in Høgh-Olesen, Henrik (ed.), *Human morality and Sociality: Evolutionary and Comparative Perspectives* (Palgrave Macmillan,

2009); also Tooby, John, Leda Cosmides and Michael E. Price, 'Cognitive Adaptations for n-person Exchange: The Evolutionary Roots of Organizational Behaviour' (*Managerial and Decision Economics*, 2006)

Trivers, Robert, *Deceit and Self-deception: Fooling Yourself the Better to Fool Others* (Penguin, 2014)

Tybur, Joshua M. et al., 'Parasite Stress and Pathogen Avoidance Relate to Distinct Dimensions of Political Ideology Across 30 Nations' (*PNAS*, 2016)

Vaes, Jeroen and Maria Paola Paladino, 'The Uniquely Human Content of Stereotypes' (*Group Processes & Intergroup Relations*, 2009)

Whiten, Andrew, 'Cultural Evolution in Animals' (*Annual Review of Ecology, Evolution and Systematics*, 2019); see also 'A Second Inheritance System: The Extension of Biology Through Culture' (*Interface Focus*, 2017)

Wolpe, Paul Root, 'Ahead of Our Time: Why Head Transplantation Is Ethically Unsupportable' (*AJOB Neuroscience*, 2017)

For the study involving Sarah Worsley, see Heine, D. et al., 'Chemical Warfare Between Leafcutter Ant Symbionts and a Co-evolved Pathogen' (*Nature Communications*, 2018)

Wrangham, Richard, 'Two Types of Aggression in Human Evolution' (*PNAS*, 2018); also *The Goodness Paradox: How Evolution Made Us More and Less Violent* (Profile Books, 2019)

The Yellowstone study was conducted by Douglas W. Smith, the senior wildlife biologist in Yellowstone National Park. There are good articles online. See Farquhar, Brodie, 'Wolf Reintroduction Changes Ecosystem in Yellowstone Park', on yellowstonepark.com (2019), accessible at www.yellowstonepark.com/things-to-do/wolf-reintroduction-changes-ecosystem

Zanette, Liana Y., Michael Clinchy and Justin P. Suraci, 'Diagnosing Predation Risk Effects on Demography: Can Measuring Physiology Provide the Means?' (*Oecologia*, 2014); also Suraci, Justin P., Michael Clinchy and Liana Zanette, 'Fear of Large Carnivores Causes a Trophic Cascade' (*Nature Communications*, 2016)

ACKNOWLEDGEMENTS

This book is a classic example of human cooperation. It has involved trust and generosity for which I'm hugely grateful. I couldn't have arrived at *How to Be Animal* without the many colleagues and friends who've given up time to share their work and offer feedback. I've been thinking about these questions for such a long time that there are lots of people I should thank, stretching back to a time before the book had begun to form. First, I'd like to thank the following individuals whose discussions have mattered over the years, in ways small and large: Jelle Atema, Francisco Ayala, Jonathan Balcombe, David Barash, Mary Barsony, Marc Bekoff, Lera Boroditsky, Ed Boyden, Michael Chaffin, George Church, Nicky Clayton, Jon Entine, Andy Gardner, Ann Gauger, Stefan Helmreich, Derek Hodgson, Nicholas Humphrey, Brian Hynek, Chris Impey, Alan Jasanoff, Bruce Jakosky, Natalie Kofler, Maria Kronfeldner, Nour Kteily, Kevin Laland, Joseph LeDoux, Tim Lewens, Rob Lillis, Greg Litus, Rodolfo Llinás, David Liu, Garret Moddel, Pascal Molenberghs, Cristina Moya, Allyson Muotri, Charles Musiba, Jack O'Burns, Thomas Pyszczynski, Tim Robbins, Michael Russell, Sophia Roosth, Jeff Sebo, Gary Steiner, Ian Tattersall, Herb Terrace, Jeroen Vaes and Robert Zubrin.

There are several fellowships and associations from which I've benefited: AHRC Centre for the Evolution of Cultural

Diversity at University College London, guided by James Steele; the Hannah Arendt Center at Bard College, with the support of Roger Berkowitz; the Durham University Philosophy Department visiting fellowship, enabled by Simon James; and, most recently, a wonderful stay at the Hastings Center for Bioethics in New York, with special thanks to Greg Kaebnick and Erik Parens for their moral support and enthusiasm. I must also thank my colleagues at the Nuffield Council on Bioethics for welcoming me, and for many inspiriting chats about all things biotechnology.

Once the book took shape, there are individuals who read early drafts and whose advice has been immensely useful to me, including: Dave Archard, John Dupre, Peter Godfrey-Smith, Andy Greenfield, Eric Olson, Michael Malay, Eileen Crist and Michael Nelson. My editors, Simon Thorogood, Paul Slovak and Nicholas Garrison, have really gone the extra mile. My copy editor Lorraine McCann was amazingly thorough and saved me from many an embarrassment. Thank you all so much. My agent Jessica Woollard has stuck with me through thick and thin. I'm incredibly grateful for her support.

But there are also several individuals who've made a special difference to me, whether they know it or not. As an untrained philosopher, I suffered from moments of considerable doubt about straying into territories for which I had no formal background. The encouragement and advice of the British philosophers Simon James and Helen Steward, and the American historian Harriet Ritvo, have cut through the anxiety. I can't thank you all enough. I must also give credit to the late philosopher Mary Midgley, who supported the ideas behind this book when I first reached out to her many years ago and gave short shrift to my concerns over a non-philosophy background. I went to see her two weeks before she died in 2018, shared cups of tea, chocolate and ginger biscuits and a passion for making sense of this strange world of ours. She was as bright and bountiful a mind as ever.

I must especially thank my friends Sam Clayton and Jeanne Greco for being a home from home. Through wine, caponata and humour, they have made my life richer. But the final tip of the hat must go to my family. I stopped writing for a while to raise my two sons, Gabriel and Samuel. But, as any at-home parent knows, the thinking doesn't stop. In truth, no education on the human condition is stronger than that offered from a child to his or her parent. But, once I started the work for *How to Be Animal*, I had to rely on the assistance of my wider family. I must thank my sister Tamsyn for stepping in on many occasions. My husband Ewan braved so many early mornings to let me write. Our conversations, arguments and shared interests have found their way into this book. I'm very lucky to have such support. I must also thank my two boys for putting up with all sorts of animal-themed shenanigans and making me cups of chai in the morning. But my deepest gratitude must go to my mother and father for a lifetime of kindness and encouragement.

LIST OF ILLUSTRATIONS

INDEX